LIFE LESSONS OF A FAILURE ANALYST

Mac Louthan

ASM International® • Materials Park, Ohio 44073-0002
asminternational.org

First printing, March 2016

Comments, criticisms, and suggestions are invited, and should be forwarded to ASM International.

Prepared under the direction of the ASM International Technical Book Committee (2015–2016), Y. Zayna Connor, Chair.

ASM International staff who worked on this project include Scott Henry, Director, Content and Knowledge-Based Solutions; Karen Marken, Senior Managing Editor; Ann Britton, Associate Product Manager; Madrid Tramble, Manager of Production; Kate Fornadel, Senior Production Coordinator; and Patricia Conti, Production Coordinator.

Library of Congress Control Number: 2015959453

ISBN-13: 978-1-62708-110-8
EISBN: 978-1-62708-111-5
SAN: 204-7586

ASM International®
Materials Park, OH 44073-0002
asminternational.org

Printed in the United States of America

To my wife Fran, our children, and especially our grandchildren, who have brightened our lives and provided the inspiration for many of the stories in this book and for many stories yet to be told.

CONTENTS

FOREWORD

I was a first-year high school math teacher, newly married, just beginning my professional career in Augusta, Ga. My dad, Mac Louthan, was speaking at a civic gathering in North Augusta, S.C., practically the midpoint between Augusta and Aiken, S.C., where he lived. The title of his talk that night was "Why Stuff Falls Apart." It was the first time I heard the lecture, an hour-long talk on ethics viewed through the lens of failures of various types and degrees of severity. Dad said something in the introduction of his talk that startled me. He was describing how he had come to be a Ph.D. metallurgist, research scientist, and college professor with a specialization in failure analysis. In other words, he was trying to answer the question, "Why is Mac Louthan the authority on 'Why Stuff Falls Apart'?" He described growing up in Bluefield, Va., and his parents' efforts to help him find direction in life as he prepared to enter college at Virginia Polytechnic Institute (now known as Virginia Tech) in the fall of 1956:

"What major should I choose in college?" Dad recalled asking.

"Well, Dickie [Mac was not always known as Mac!], you should study something you are good at," my grandparents replied.

My dad then followed a familiar path of self-deprecation as he addressed the audience: "The thing I was best at was failure, so I decided to become a failure analyst."

Failure? This was not what I envisioned as my father's greatest skill! In fact, failure was something I did not associate with Dad at all. I loved the stories of his high school basketball team winning the Virginia State basketball championship in 1956. I knew my dad could put us onto fish whether we were in a lake, on the river, or in the surf in the Gulf of Mexico. With great pride I had followed my dad down the steps at Cassel Coliseum for our front row seats behind the VPI bench (Dad still won't legitimize the name Virginia Tech) with tickets we possessed because of Dad's former playing days. I remember going to graduations at Virginia Tech—I mean VPI—as Dad was recognized many times for excellence in teaching. I was afforded many opportunities to travel across the United States, and even into Singapore, because Dad was sought out as an engineering consultant and conference speaker. Dad was not best at failure, though he was apparently very good at analyzing it.

If not failure, then just what was Dad best at? I have come to realize over the years, even decades now, since that night in North Augusta, S.C., when a Ph.D. research scientist from Savannah River National Laboratory spoke to a Rotary Club audience that included local businessmen, workers, and at least one high school teacher and coach, the answer to that question is probably revealed in the fact that Dad was asked to give that lecture to that audience in the first place. He has delivered the same speech hundreds of times to numerous types and sizes of audiences ever since. "Why Stuff Falls Apart" has been heard by audiences all over the world, from large auditoriums to high school classrooms, to corporations and colleges, to scholars and athletic teams, because Dad has a unique way of making very complex ideas and systems—and yes, failures—understandable. Dad demonstrates engineering principles, not merely through hard-to-read textbooks and formulae, but also by twisting carrot sticks until they break or pulling apart Snickers bars. Dad sees engineering in daily life and is able to bridge the intellectual gap and make it understandable.

Dad never used "Why Stuff Falls Apart" to amaze people with his knowledge. He used it instead to say that the things that tend to create failure in bridges, crane arms, drive shafts, and other engineered systems composed of engineered materials are the same as the things that cause failures in careers, marriages, churches, and other human relationships. Stuff falls apart because people, not just engineers, tend not to do what we ought to do—the heart of ethics. Dad understands this. Dad needed a Ph.D. to study some failures that are visible only to an electron microscope, but he has a unique way of taking those same principles and seeing how they are repeated in our daily experience.

Raising three children in a disciplined household gave Dad plenty of case studies in failure. Spoiling nine grandchildren as he erodes the whole concept of discipline in an act of revenge against our childhood rebellion has given Dad countless anecdotes and stories. He has editorialized his family and friends repeatedly in technical journals and lectures. Why? Because his ability to relate, and relate to almost anyone, is Dad's greatest skill.

In the pages that follow, you will read stories that do in fact relate to the science of engineering and the study of failure. But you will find them intertwined with names and life experiences of people Mac Louthan loves dearly and very well. Some of the experiences have been funny from the start. Some have been very difficult, even in the memory of them. All of the anecdotes relate to some engineering principle, ethical idea, or both. But the engineering principles and ethical ideas are not limited in their application to the industries and sciences they support. So Dad weaves them all together. For Dad was never able, nor willing, to separate his observations about the life he experienced with his family from the systems he analyzed in the laboratory. May we all learn the lesson of living such an integrated life, a life of integrity, like my dad, Mac Louthan.

Keith Louthan

PREFACE

Many people consider engineers and scientists to be technical geeks involved in academic exercises designed to satisfy intellectual curiosity. This opinion contrasts the self-image of most practitioners working in these arenas. We generally consider ourselves to be family members working for the good of our corporation, state, and nation. Engineers and scientists are active leaders in community and church activities, activate the moral compass of professional organizations and nonprofits, and provide a positive influence on youth as well as the elderly. Unfortunately, this more human side of the engineering and scientific community often goes unnoticed and unrecognized. This lack of recognition deflects the attention of service-oriented youth toward other professions and creates an ever-increasing need for technically educated professionals.

The stories collected in this book were written over a twelve-year period as part of an editorial series published in the journal originally named Practical Failure Analysis (2001 to 2003) then continued as Journal of Failure Analysis and Prevention (2004 to present). The goal of the editorials was to use examples from everyday life to illustrate how family and friends, recreation and responsibility, work and leisure tie together to make a person. Written to influence the failure analysis community, it became apparent over time that the lessons in most editorials had broad applicability. The importance of study, hard work, goal setting, and practice are demonstrated, as is the influence of integrity on success and satisfaction. The editorials revolve around observations made in classrooms, lunchrooms, gyms, highways and back roads. Lessons are taught by children, grandchildren, friends, and acquaintances. This collection illustrates that professional activities cannot be isolated. Life's lessons surround us if we will only recognize the teachers. Engineers and scientists may be technical geeks, but their professional and social interactions play a major role in determining the quality of our lives.

ACKNOWLEDGMENTS

This book resulted from the trust my family, friends, and ASM International placed in my storytelling ability. Mary Anne Fleming, senior content developer, journals, at ASM International, continually provided encouragement and support throughout my tenure as editor of the Journal of Failure Analysis and Prevention. My wife Fran actually encouraged me to short cut family time to write the editorials and often edited a story before it was published. Friends and family served as examples for many of the stories, and occasionally a story was referenced or given to a group of students. Several colleagues suggested that the stories be published as a collection. However, this publication is due, almost solely, to the efforts of Ann Britton, associate product manager at ASM International. When the decision to publish was reached, Ann took charge and carried the project to completion. Publication of this collection of stories would not have occurred without Ann's insight and efforts.

All of the editorials included in this publication have been republished from Practical Failure Analysis (2001 to 2003) and Journal of Failure Analysis and Prevention (2004 to 2011) with kind permission from Springer Nature.

ABOUT THE AUTHOR

Mac Louthan is a retired senior consulting scientist from Savannah River National Laboratory, where he spent over 30 years working in research and management. He taught metallurgy and engineering science at the University of Notre Dame and Virginia Tech, where he was elected to the Academy of Teaching Excellence and received the Sporn (student-elected) and Wine (faculty-elected) Awards for teaching excellence. He is an ASM International fellow, past president of the International Metallographic Society, on the board of trustees of the National Youth Science Foundation, founding editor-in-chief of Practical Failure Analysis/Journal of Failure Analysis and Prevention (2001 to 2011), and past member and chairman of the National Nuclear Security Administration's Network of Senior Scientists and Engineers. The author also served as organizer and/or chairman of eight major technical conferences, was on the editorial review board of several peer-reviewed journals, and edited five volumes of Microstructural Science. He has given invited lectures throughout the United States, Canada, Europe, Asia, and Australia, published over 200 peer-reviewed papers, edited seven books, received several Best Paper awards, and presented conference keynote addresses to various professional societies. His professional awards include: the Orth Award (SRS), several Westinghouse Signature Awards for Excellence, the Sorby Award (IMS), Distinguished Life Member of Alpha Sigma Mu, the Putman Award for Service to ASM International, the President's Award from IMS, election as a Distinguished Educator by the Materials Engineering Institute, a Certificate of Achievement for a distinguished career in Hydrogen Effects on Materials, and election as a Distinguished Scientist by Citizens for Nuclear Technology Awareness. Mac has coached Little League and college sports, held leadership and teaching positions in his church, been active in high school booster clubs in Virginia and South Carolina, was a founding member of the University of South Carolina Aiken's Pacer Club, and is most proud of Fran, his wife of 54 years, his three children, their spouses, and his nine grandchildren.

INTRODUCTION—FUNDAMENTAL CAUSES OF FAILURE

Failures do not just happen, they are caused by the actions or lack of actions by people. A good failure analysis will determine the cause of a failure and establish preventive measures to mitigate the potential for similar failures in the future. The six fundamental causes of failure are:

- Improper design
- Poor materials selection
- Defects in materials
- Improper processing
- Errors in assembly
- Inadequate service

These causes are referenced in many of the stories in this book and readers are asked to reflect on how continual effort to avoid the failure causes can lead to success in life's endeavors.

THE POWER OF WORDS

ANSWERING THE QUESTION WHY

My grandchildren call me Big Mac, and during family get-togethers over the Christmas/New Year's holidays, there was one question that I constantly faced: "Why, Big Mac, why?" This one question was actually a combination of lots of questions: "Why did the glass break?" "Why doesn't that screw fit?" "Why did the batteries burn out?" "Why, Big Mac, why?"

Frequently, my answer to one why would simply invoke another why. Trying to explain a failure to preschool children is tough. Unlike many of us, preschoolers* are not timid about questioning a situation, especially if they perceive the situation to be important.

My son Keith and I were building a hutch for his wife Amy for Christmas. My efforts were primarily relegated to the menial tasks, but I was allowed to nail several of the shelves to the back of the hutch. Austin, age six, and Hunter, nearly three, were actively trying to help me because Keith had said, "You can help Big Mac if you want."

After I had driven several nails, Hunter asked if he could "hammer a nail." I drove three nails halfway into the hutch, handed Hunter the hammer, and watched as he attempted to drive the nails the rest of the way. He successfully drove the first two nails, bent the third nail, and immediately asked, "Why did that nail bend?"

The teacher in me saw the question as an opportunity to expose an interested child to materials engineering. I talked about mechanical properties, bending, and may have even

---

* I continually use spell checker when writing. When I first wrote *preschoolers* in this editorial, the word was underlined in red. One recommendation for a correct spelling of *schooler* was *scholar*. I wonder if the willingness to ask the question *why* isn't a major factor in the transformation of a preschooler into a scholar.

mentioned yield strength. Austin interrupted my discourse by stating, "He just hit the nail wrong, Big Mac."

Austin had effectively described the root cause of failure by finding a simple answer to Hunter's question. I had missed the mark by attempting to include way too much information.

Experience has shown that many engineers and analysts fail to effectively communicate because they overwhelm the audience with details and facts. In some cases, the facts are basically irrelevant and the details hide more than they reveal. Austin had it right. Hunter needed to know that the nail bent because he hit it wrong; any other fact was irrelevant to the failure analysis.

A key element to the successful communication of the results of any failure analysis is to answer the question *why* as simply and clearly as possible. Frequently, we hesitate to give the simple, direct solution because we want to communicate the extensive effort and vast knowledge base required to develop the answer.

The statement, "He just hit the nail wrong," doesn't convey the message that an expert had to work long and hard to answer the question. The effective transfer of information from a failure analysis expert to a designer, fabricator, and/or user requires that the answer to the question *why* be understood. Additionally, the understanding of why must show how that information can be applied to improve the product performance, increase system lifetime, and/or provide for cost-effective changes in operations.

Communication is not a one-way street. As the failure analyst attempts to provide simple direct answers to the questions raised by a failure, the designer, fabricator, and user must continually probe with the question *why.* The question must be asked over and over again until the answer is plain and fully apparent.

My Dad once told me, "If the answer to a problem is complicated, you don't understand the problem." I'm convinced that what Dad said is true. My grandchildren are now demonstrating that you develop understanding by never hesitating to ask why.

Practical Failure Analysis, 2001, Vol 1(2), p 3

DEFECTS, T-SHIRTS, AND FAILURE ANALYSTS

The word *defect* is defined as the "lack of something necessary for completeness" (*Webster's New World Dictionary of the American Language*). This word takes on several meanings in the metallurgical/materials community. Vacancies, dislocations, grain boundaries, and even external surfaces are defects in a crystal lattice. Alloy atoms that were intentionally added to metal exist as substitutional or interstitial crystalline defects in the base metal lattice. Inclusions, porosity, precipitates, and segregation are defects that are readily observable by optical metallography. Scratches, stains, and regions of roughness are defects on polished surfaces. Fatigue failures initiate at a variety of defects, including defects originally present in the material, service-induced defects, and mechanical and/or design-induced defects. Inclusions, pores, pits, wear marks, regions of intergranular attack, holes, notches, sharp corners, and even machine marks may act as sites for failure initiation. The failure analyst must be careful with the use of the word *defect* and should thoughtfully consider the definition in the *Metals Handbook Desk Edition* (2nd ed., 1998): a defect is "a discontinuity whose size, shape, orientation or location makes it detrimental to the useful service of the part in which it occurs," or it may be something that "renders a part unable to meet minimum applicable acceptance standards or specifications."

My favorite T-shirt is red and decorated with a fish, the words *Big Mac,* and the initials *KL.* The shirt is two years old and was decorated by my now seven-year-old granddaughter, Kate Lilly. Although the red has faded and some of the decorations have worn and washed off and the lettering does not meet drafting standards, I question the existence of defects in either the lettering or the drawing of the fish. In fact, the lettering characteristics that a draftsman would consider defective help make the shirt my favorite.

Every engineered component contains something that will be termed a defect by some branch of materials engineering and science.

Additionally, although the pelvic, pectoral, and dorsal fins and the colors of the fish do not match those of any known species and would be termed defective by an ichthyologist, it is the uniqueness of the fish that draws attention to the shirt. On several occasions someone standing near me has looked at the shirt and remarked, "That shirt must have been made by someone very special." For two years the shirt has functioned properly by covering my torso, keeping me warm, and allowing me to enter places where "shoes and shirts are required." Furthermore, the shirt has brought joy to my life and allowed me to show my pride in Kate's artistic abilities. Defects? I don't think so!

The materials literature contains numerous references to structural and microstructural inhomogeneities that could be termed defects and exist in virtually every component. A cold worked component may have a dislocation density exceeding 10^{10} centimeters of dislocation line per cubic centimeter of material. However, even though dislocations are defects in a crystal lattice, they are not defects to the cold worked component. They are microstructural characteristics necessary to obtain the specified materials properties. Aluminum oxide inclusions are defects in the optical microstructure but are not generally considered defects in the sheet, plate, foil, tubing, or wire products made from an 1100 aluminum alloy. Every engineered component contains something that will be termed a defect by some branch of materials engineering and science. However, the failure analysis community must try to avoid calling something a defect unless that something played a role in the failure process, or, by its very existence, that something would cause the component to fail to meet specifications and/or be rejected Additionally, we must be aware that the term *defect* is often used to describe common microstructural inhomogeneities in a material, although the inhomogeneities are normal and may have been required for component fabrication. Wrought metal products are frequently purchased to a grain size specification to assure the presence of sufficient grain boundaries (crystalline defects) and a maximum average grain size. Cast products generally contain segregation and porosity (microstructural defects) and inclusions; second-phase precipitates and other microstructural defects are present in most structural metals and alloys. Clearly, defects in the crystal lattice and microstructure are not necessarily defects in the component. However, when the public hears that defects are present in a component, regardless of the size or character of the defect, the public will generally think that something is wrong and the manufacturer is responsible. Hence, we must be careful with our choice of terms or our words may be misinterpreted.

Kate and her brother Tripp gave me another T-shirt this past Father's Day. The commercially decorated shirt is beautifully lettered with the words "World's Best Grandpa." I was wearing that shirt in a checkout line at Wade's Grocery in Radford, Va. The clerk was in her mid- to late teens, and, after glancing at the shirt, she smiled and asked, "How did you get my Grandpa's shirt?"

The materials engineering and science community wears many T-shirts: T-shirts for professors, for producers, for designers, for materials selectors, for inspectors, for metallographers, and T-shirts for failure analysts. Unless you are careful with the use of the word *defect,* you could cause a very public problem and might even be accused of wearing someone else's shirt.

Journal of Failure Analysis and Prevention, 2004
Vol 4(6), p 3–4

CRYING WOLF

Years ago, our first son Rick loved the old Russian folk tale "Peter and the Wolf" and would play a recording of the story over and over. On one side of the record was a storyteller's narration accompanied by classical music. The other side of the record played the music without the story. Rick didn't seem to care which side of the record he heard, but even when he was listening to the music-only side he was hearing the story in his mind. He would frequently say, "Here comes the real wolf!" just as the music signaled the wolf's approach. I was amazed that a very young child developed such a strong association between the story and the music. However, life's lessons have shown that most of us develop associations, and often the associations strongly influence our perception of reality.

Recently my wife and I were driving north on route 1-77 through North Carolina. We have made this trip often, and virtually every trip involves passing through a stretch of highway marked as if there was road work being done. However, the truth is we generally see signs stating that there is Road Work Ahead, but we seldom see any road work. During this past trip we drove approximately six miles where the right lane was closed, presumably for road work, but there were no signs of work in progress. Because this experience has been all too common, we began to discuss how highway signs frequently "cry wolf." We have seen signs stating Dense Fog Ahead when there is no fog to be seen, signs calling attention to Ice on the Road when the temperature is far above freezing, and signs posting a speed limit of 55 mph while nearby vehicles are traveling at 70 mph. Those signs are simply crying wolf! Accidents often happen near such signs when the wolf actually appears. How many times do we have to pass a sign that wrongly alerts for a work zone, ice, high winds, or fog before we begin to associate the sign with misinformation and totally ignore its warning? I'm convinced that if wolf is called when there is no wolf around, eventually everyone will ignore the call. Wholesale ignoring of a wolf call is one reason we have multicar/truck pileups in dense fog even when fog alerts are posted along the roadside.

As failure analysts, we are frequently charged with the task of calling an alert to a potentially dangerous system, component, or situation. How, when, and where we make that call may be as critical as the call itself. We certainly don't want to cry wolf every time we see a failure. Some things simply wear out; few, if any, things are perfect 100% of the time; and tragedies don't always occur just because a scenario for the tragedy can be mentally developed. A manufacturer is not unethical just because one of the several million "do-whats" that were produced over a ten-year period failed, even if the failure resulted in an injury. We need to carefully consider the results and general implications of cries that are associated with our failure analyses. Before we decide to cry wolf, let's be certain that the wolf is present. We clearly don't want our reports to become highway signs that cry wolf so often that our cry is generally ignored.

Rick associated the music on his record with the story "Peter and the Wolf," and, unfortunately, I often associate highway signs with incorrect information. What do our readers associate with the failure analysis reports we produce: excellence, insight and thoroughness, thoughtful reflection and appropriate recommendations, or a cry that the wolf is knocking at the door? Have you ever been a "Chicken Little" and predicted that the sky is falling, even when you knew it really wasn't? A willingness to cry wolf and/or predict disaster may provide attention over the short run, but an excellence in prediction must be based on the insight gained through thorough evaluation and knowledgeable reflection before it can provide a solid foundation for future work. Our reputation is the background music for most of the work we do. What associations does that music bring to potential customers?

Folk tales frequently use a simple story to describe undesirable behavior. Let's be certain that our reports, our analyses, and our expertise do not cry wolf unless the wolf is really at the door.

Journal of Failure Analysis and Prevention, 2005
Vol 5(5), p 3

PREPARING THE CUSTOMER

Failure analysts and metallurgists use a language that is foreign to most people. The technical meanings of terms such as *beach marks, cleavage,* and *necking* differ greatly from their meanings in normal usage. Other technical terms such as *martensite, pearlite,* and *2024-T6 aluminum* aren't even in the layman's vocabulary. Preparing the customer to understand "the language of metallurgy" is, therefore, a critical factor in the success of any metallurgical/failure analysis presentation.

We recently visited Nancy and Bob Rappleyea, family friends who live in Florida. It had been 20 years since we had vacationed together on the west coast beaches of central Florida, and we were astounded to learn the current value of beachfront property. Two-room shacks that we had previously rented for less than a hundred dollars a week had been transformed into million-dollar vacation cottages by the addition of a coat of paint, a new name, and a carefully painted sign. Apparently, vacationing in a place called Seascape is worth much more than vacationing in the same place if it is called Joe's Rentals. The realtor readily recognizes the value of fresh paint, flowery language, and a handcrafted sign.

During our earlier vacations in Florida, our children played with Bob and Nancy's children and with the children of Karen Greer and other mutual friends. The children are now grown and many have families of their own. Karen's daughter Tammy has a husband, two boys, and a bunch of horses. Tammy and Rodney, her husband, recently purchased some land (for the horses) and informed the children that they were going to build a house out in the country. However, as any horse person will recognize, a barn came before the house. When the barn was completed, Tammy and Rodney took the children to see the finished product. The children's immediate reaction was surprising. The

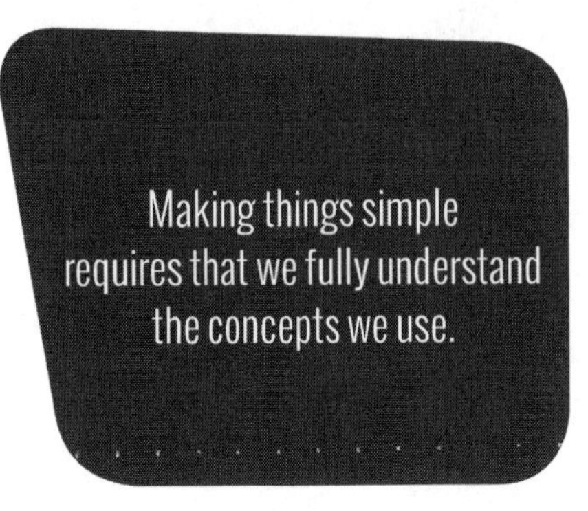

oldest child stated, "I don't like it," while the youngest simply looked at the stalls and asked, "Which one is mine?" Neither child had been prepared to visit a barn, and neither child understood what they were seeing. The language at home had focused on a new house and that is what the children had expected to visit. All too often the language of metallurgy doesn't prepare the customer for our presentations any better than Tammy had prepared her children for the visit to the barn.

Preparing the customer often requires education. Unfortunately, the first person we have to educate is frequently ourselves. We use terms and descriptions that we don't completely grasp. My daddy used to tell me, "If you can't make it simple, you don't understand the concept." Clearly there are exceptions to this rule: Einstein, Fermi, Newton, and Franklin, for example. However, most of us are far from Einsteins, and even if we did approach him on the intellectual level, we would have to make it simple for our customers. Making things simple requires that we fully understand the concepts we use.

Most of us recognize that martensite, in iron-base alloys, is formed by the shear-induced transformation of austenite to the metastable, body-centered tetragonal phase that we term martensite. We also understand that the martensite transformation takes place when processing conditions do not provide time for diffusion and the formation of ferrite and cementite. A few of us will be surprised to learn that an alloy whose structure is tempered martensite might not contain any martensite in the microstructure. The tempering operation converted the martensite into a ferritic matrix that contains nanoparticles of cementite. These cementite particles are generally too small to be resolved in an optical microscope, and an optical micrograph of tempered martensite will look very similar to that of as-quenched martensite. If we are trying to convince a customer that a failure occurred because the martensite was not properly tempered, we need to be able to communicate, among other things, what the tempering operation did to the microstructure, how the microstructure affected the mechanical behavior of the component, and why the improper tempering caused the problem. We can't explain the importance of tempered martensite if we don't understand what the tempering really does, and if we can't communicate martensite, we may not be able to prepare the customer to receive and understand our presentations.

There are two primary reasons that we fail to prepare the customer to fully appreciate our presentations:

- We use a language (terms and acronyms) that is foreign to, and not understood by, the customer.

- We don't fully understand the concepts associated with the failure and cannot make them simple enough for the customer to understand.

I believe that, like beach property in Florida, our presentations can be made more valuable if we paint the report by interpreting our language and being sure that the customer understands the terms and concepts. Additionally, we should take the time to properly name the presentation and use the tools we have to prepare signs that lead the customer to the proper conclusion. If we don't, the customer may think we are describing a house when we are presenting a barn and then ask, "Which stall is mine?"

Journal of Failure Analysis and Prevention, 2006
Vol 6(4), p 3–4

GOVERNMENT WASTE, BANDWAGONS, AND A NEED FOR SHARING

The governor of the commonwealth of Virginia recently bet the governor of Kansas that Virginia Tech would beat Kansas in the Orange Bowl. The wager included the promise to send the Kansas governor a Virginia ham if Kansas won. On the same day that I heard that the Virginia governor was using pork to lure the Kansas governor into a wager, I also received a report from Congressman Rick Boucher informing his constituency that he had obtained federal grants, better termed *pork,* for the town of Pulaski.

There are lots of empty buildings in Pulaski, primarily because the furniture industry has exited the area and was shut down in favor of overseas operations. The pork that Congressman Boucher discussed was in the form of two grants in the total amount of $272,500 for "the planning of a nanotechnology business park in the town of Pulaski." I live in Pulaski County, go to church in the town of Pulaski, and would love to see the empty buildings transformed into vibrant commercial centers... but a nanotechnology park? Congressman Boucher is jumping on the nanotechnology bandwagon, and because of his jump, over a quarter million dollars will be used to provide "an analysis of the infrastructure in Pulaski, as well as a study of existing structures."

Virginia politicians are no different than the politicians from other areas, with pork being part of their daily diet. They gather pork, swap pork with other politicians, serve pork to the public, and even used government funds to notify their constituency that pork is on the way. The notice I received from the Congressman said, "This mailing was prepared, published, and mailed at taxpayers' expense." A constant pork diet is not healthy! Stuffing pork into depressed regions almost always results in failure, and as failure analysts we should call for a change. However, before we call for a change in other organizations, we must be willing to experience change ourselves.

Each year there are numerous conferences, seminars, and symposia that have the theme of failure analysis and prevention. Each year numerous presentations at these public forums will go unpublished, and each year many presentations will be repeated at several forums. We love to talk, argue, and demonstrate technical expertise but often find it difficult to do the work necessary to actually develop a paper and provide a finished, well-referenced technical report to our peers. Just as the politicians love to serve pork and jump on the latest technology bandwagon, failure analysts and many other scientists and engineers often serve "half-baked" work by jumping on the public forum bandwagon. It is easy to make a presentation but hard to provide the supporting documentation that can transform a presentation into a document with lasting value. It is easy to modify a presentation and stand on another stage presenting essentially the same work a second or even a third time. However, it is hard to join with our peers, remake the presentation, and share a document that presents the fundamental scientific underpinning that transforms the presentation into a document that actually educates the reader.

Politicians are well known for taking the easy road and serving pork sandwiches but floundering when real progress is necessary. I am concerned that the rest of the world is not much better. The town of Pulaski will munch on the nanotechnology sandwich and, most likely, realize that the sandwich would be better with some mustard, lettuce, and tomato, and ask for another government grant to provide the condiments. It is unlikely that a successful nanotechnology park will ever exist in Pulaski, because virtually every politician is serving similar pork to their constituents, and the country needs only a few nanotechnology parks that industry will provide to areas where success is best served. I am always amazed that government taxes industry, sifts the tax dollars through a costly bureaucratic process, and then uses the leftover tax dollars to entice that same industry to relocate to technology parks that are fragmented throughout the country. However, I should not be amazed: I have done the same thing and perhaps, just perhaps, so have you.

There is an old adage that says, "I'd rather see a sermon than hear one any day." Technical presentations are sermons that we hear, while documents are sermons that we can see, touch, and even study. The next time you make a technical presentation, think about those politicians that send hams, serve pork, and always move on a well-paved path but never climb the mountains of significant concern. If your presentation is important enough to serve, is it not important enough to package correctly so that it can be shared with the science and engineering community and not simply given as pork to those who happen to hear the presentation?

Journal of Failure Analysis and Prevention, 2008
Vol 8, p 91–92

INFLUENCING OR INFORMING?

The packet that arrived in the mail in April contained a car key and a slick paper foldout advertisement stating that I had won the car or $7000 or one of a group of prizes that included a Rolex watch, an MP-3 player, and a $100 bill. To receive the prize already designated for us, we had to visit the car dealership, try the key in the car door, and if the key didn't work, simply claim the other prize. If the key worked, we could drive off in a new car. I knew that Kate, our 10-year-old granddaughter, would be visiting us during the "try the key time" listed in the advertisement and thought that she would enjoy participating in the quest for our prize. I told Kate that if we won the car or the $7000, I'd keep the prize, but if we won anything else, the prize would belong to her. I was surprised at her response: "It's a sham, Big Mac—we aren't going to win anything." I showed her the advertisement and the statement that we had definitely won one of the prizes listed, and she simply stated, "I'll go with you if we can take the truck, but they are just trying to get us to come to the car store and we aren't going to win anything."

We took the truck, and by the time we had reached the dealership, Kate was a little more enthusiastic. I tried to convince her that because of U.S. laws related to advertisements sent through the mail, we could be relatively certain we would receive one of the prizes. Kate recognized the MP-3 player shown in the advertisement and assured me that the player cost more than $100, so she was sure that she would be receiving the $100 bill or, if she was lucky, one of the other prizes. As we pulled into the car lot, there was a line of people, keys in hand, waiting to talk to someone. Kate jumped out of the truck and got in line while I parked. When our turn arrived, Kate couldn't unlock the car so we got to talk to a sales agent. "I told you it was a sham" was the only comment that Kate made.

The sales agent wanted to discuss our purchasing a new car, but Kate said, "We just want our prize," so the agent entered our prize number in a computer and did a search for the prize we had won. Kate began to get excited when the search revealed that we had won the MP-3 player. The sales agent went into a "back room" and returned with a certificate that stated we would receive the MP-3 player if we sent a certified check for about $22 (for shipping and handling) to the address listed on the form and also informed us that the player might differ from the player shown in the advertisement. As we drove out of the lot, Kate said, "I knew it was a sham. I just knew it." We have sent in the certified check, but as of late May we haven't received the MP-3 player—not even the cheap version that wasn't pictured in the advertisement. Kate was here this past weekend and is still convinced that the advertisement was a sham and that the information in the ad was carefully prepared to influence us to come to the dealership rather than inform us about the prize and the prize claiming process. I keep hoping that the prize will arrive and that Kate will ultimately enjoy the experience rather than simply adding to her knowledge that most advertisements are designed to influence rather inform.

This experience brings up several questions that might apply to the failure analysis process: How are our failure analysis reports designed? Do the reports use slick paper to lure the reader down a path that hasn't been well prepared? And, are the reports more involved in the process of influencing than in the process of informing?

There is a subtle difference between influencing and informing, and most of the time it is the things that are not said that contribute to the difference. The news media often distort the news by leaving out many of the relevant facts. I've seen photographs of police officers using force to subdue a criminal under a headline stating "Police Brutality," while the report leaves out any discussion of the five previous minutes when the criminal was swinging a club at the police officer. The objective of the photograph and the discussion in the paper was not to present the news but to influence the news reader.

I was once interviewed by a reporter from *The New York Times* concerning some stress-corrosion cracks in a brace that had been long removed from a nuclear materials production reactor. The reporter wasn't very interested in the history of the brace but wanted to discuss the cracks. He even wanted a scanning electron micrograph (SEM) of the crack that was magnified to suggest that the crack was about a half-inch wide, when the actual width was only several microns and the crack did not penetrate the wall thickness of the brace. The reporter was attempting to influence his readers and had no apparent concern about presenting information surrounding the brace. All he needed to do was publish the SEM photograph, leaving out the magnification, and his story would sell itself. Fortunately he was not given the photograph. Listen to a news report concerning an event where you have been personally involved and see how the story can be distorted simply by leaving out certain facts related to the event.

As failure analysts, we certainly don't want to distort our stories by omission—at least I hope we don't—even though our jobs are such that in many cases we must try to influence a customer, a judge, or a jury. However, as we try to influence the client, let's be certain that we fully inform and don't distort the story by skipping over the part of the analysis that doesn't support our case.

The next time you are preparing a report, remember the key, the slick paper, and a ten-year-old girl's cry that "I knew it was a sham. I just knew it!" Certainly we aren't dealing with children, but unless we offer a complete analysis, aren't we becoming too much like a salesman attempting to influence without informing?

Journal of Failure Analysis and Prevention, 2008
Vol 8, p 309–310

IN PURSUIT OF EXCELLENCE

REFLECTIONS OF THE SUMMER

Summer is my favorite season; at least it is in my top three. There are lots of reasons for summer to be a favorite season—the weather, vacations, family birthdays, fishing, and camp. Even though I'm over 60, camps are still among my favorite summer activities. I've been working with the National Youth Science Camp for about 20 years and have worked with ASM International's Materials Camp since its inception two years ago. Both of these camps are for highly qualified students and their cost for participation is minimal: the camps cover the cost of airfare and provide local transportation, room, and board. Materials Camp focuses on metallurgy and failure analysis, while science and outdoor activities are the focus of the National Youth Science Camp. At both camps, the campers'/delegates' enthusiasm is contagious, and after a few moments of association with these very gifted and extremely talented campers, even an old cynic sees hope for tomorrow.

Materials Camp is supported by the ASM International Foundation. The National Youth Science Foundation and the state of West Virginia support the National Youth Science Camp. The governor of West Virginia requests that each of the 50 states select two of the top science students from that year's high school graduates. The selected students become delegates to the month-long science camp. Distinguished engineers and scientists from throughout the United States present lectures and/or provide directed studies to the delegates. Outdoor activities include rock climbing, spelunking, kayaking, hiking, and camping. Although most of the activities of the National Youth Science Camp are held near Bartow, W. Va., a trip to Washington, D.C., and a Senate luncheon are also included on the agenda. I present the lecture "Why Stuff Falls Apart" to both camps.

This lecture discusses the responsibility that engineers and scientists have to society and illustrates the importance of ethical behavior in failure prevention. Virtually every time the lecture is presented, several delegates hang around after the lecture to discuss a disappointing encounter with an engineer or scientist. Apparently, engineering activities, even those of metallurgical engineers, are frequently viewed as exploitation rather than extroversion. Many of the delegates are surprised to hear an engineering professional speak of personal responsibility, and the idea that an engineer should have more interest in his environment and other people than in himself is totally foreign. The exceedingly bright, exceptionally talented young people who serve as delegates to these two camps simply do not think of engineering as a caring, concerned profession.

The delegates generally consider medicine, social work, teaching, and ministry as the caring professions. Engineering is defined in *Webster's New World Dictionary: Second College Edition* as "the science concerned with putting scientific knowledge to practical uses." Surely the practice of putting scientific knowledge to use is caring. The impact of this caring is readily apparent in the National Academy of Engineering's list of the 20 greatest engineering achievements of the 20th century. The engineering accomplishments associated with the list (electrification, automobile, airplane, water supply and distribution, electronics, radio and television, agricultural mechanization, computers, telephone, air conditioning and refrigeration, highways, spacecraft, internet, imaging, household appliances, health technologies, petroleum and petrochemical technologies, laser and fiber optics, nuclear technologies, and high-performance materials) have made our world a better place. Why haven't the engineers who made this list of accomplishments possible been widely recognized as caring people?

Engineers have developed national and international codes and standards as formalized ways to protect the public. Even the engineers working in the defense sector are working to protect the interests of their nation. Perhaps the teaming required for most major engineering achievements overshadows the contributions of any single engineer. The teaming argument may be partially true; however, even when individual engineering achievements are widely recognized, the public may not associate the achievements with caring. Franklin's and Einstein's achievements were in the public interest! Why are they not recognized as great humanitarians, polished with a love for humanity similar to that attributed to Pasteur and Salk?

When asked these questions, the delegates to both camps generally attribute their lack of awareness that engineering is a caring profession to "what they had heard." Engineering is interesting, challenging, high paying, a pathway to individual success, a boot strap profession, and for nerds or math freaks. The concept that engineers serve to protect the public interest is not shared with most high school students. Why? Could it be that the practicing engineers (you and I) don't push the public interest aspect of their job? The ASM International Foundation pamphlet "Find Out More about the Career that Shapes the Future" states that "by developing artificial skin for burn victims and chromium alloy hip implants, materials engineers are improving the quality of life for many people." This statement, however, is hidden in the section "Where Do Materials Engineers Work?" Even when we advertise the practice of engineering, we fail to emphasize the importance of protecting and serving the public. The pamphlet "Making a Difference" advertises membership in ASM International but does not illustrate how ASM International helps its membership "serve the public."

The saying "You are what you think" should apply to engineers. One glance at the list of great engineering achievements in the 20th century clearly demonstrates that the engineering profession has made significant positive contributions to society. The engineering profession is simply a collection of engineers; thus, engineers have made contributions to the public welfare. If we would simply recognize that protecting the public is a job responsibility and realize that our career is the major contribution, perhaps we could alter the public perception of engineering and have young people think of engineering as a caring profession. However, to actually accomplish this transformation, we need to demonstrate that we care.

One of my favorite stories concerns the wall that separates heaven from hell. As many of you know, St. Peter inspects the welded type 304L austenitic stainless steel wall every Thursday afternoon. It seems that the wall welders were not ASME certified, thus there is some concern over possible weld defects. St. Peter has very sensitive hearing and simply uses his hearing as an acoustic monitor of the vibration damping that accompanies his footsteps. This monitoring permits him to determine the location and size of any defects in the wall. One Thursday he discovered a defect in Section 8, panel 4, weld 127 of the beta portion of the wall. He rushed back to the office and called Satan to discuss the devil's responsibility for wall construction and repair. By celestial agreement and through numerous signed contracts, wall care and maintenance were to be provided by hell. The call ended with the devil stating that he would "get right to it." Thus assured of an immediate repair, St. Peter forgot about the wall defect until the next Thursday when he discovered that the defect not only had not been repaired but had grown to the extent that a soul could get through. A very angry St. Peter called the devil again and assured him that "unless the problem is fixed by Monday afternoon, heaven would, of necessity, take hell to celestial court." The devil's response was simply, "That is fine with me, Pete. Where will you get a lawyer?"

This story loses some of its humor if it is rewritten to emphasize heaven's need for a failure analysis and ends with the devil asking, "Where will you get an engineer to analyze the failure?" Suddenly the need to lead a caring life becomes much more personal. Instead of laughing at the lawyers, we are smiling at ourselves. This change in the story leads to two very important questions: How does my career influence the public's opinion of the engineering profession? And, more importantly, do I care enough to be certain that what I am doing will help the public?

Practical Failure Analysis, 2001, Vol 1(5), p 4–5

GIVING A DIAMOND

I was retyping and editing an editorial for this issue of *Practical Failure Analysis* on Tuesday morning, September 11, 2001, when I first heard about the terrorist attack at the World Trade Center. Since that time, I've discarded the original draft and mentally rewritten the editorial on numerous occasions. Each rewrite was an attempt to find appropriate words to express the conflicting emotions that resonate from the disaster: love and hate, concern and anger, hope and despair. The cowardly perpetrators rejected any claim to humanism, while victims, especially the first responders, demonstrated the best of human characteristics. I almost succumbed to the temptation to analyze the cause(s) of the disaster and, as any failure analyst should, to determine what must be accomplished to assure against similar failures in the future. However, the enormity of the tragedy quickly brought recognition that I'm unlikely to properly analyze the failure.

First, virtually all my information is secondhand and filtered through media moguls that are much more interested in securing an audience than in providing the truth. Facts and fiction are mixed without concern and many news reports truly belong on an editorial page. Second, I'm not qualified to interpret the information evolving from the ongoing failure analysis process, and third, even if I were able to determine the proper solution, implementation of that solution would likely require resources that are generally unavailable to individuals. In fact, I not only can't analyze the failure, I can't find words to express my feelings. Therefore, this editorial will simply describe an approach to the failure that is within virtually everyone's grasp: giving a diamond.

Several years ago, my wife asked for a diamond-like pendant for Christmas. This was going to make Christmas shopping easy because she usually fails to give any

Christmas hints. I knew that she requested "diamond-like" because she wanted a large stone and recognized our budgetary restrictions. The local jeweler stated that he didn't sell cubic zirconia and suggested checking with a department store. The department store was well aware of the diamond-like qualities of cubic zirconia and offered several beautiful, multi-carat pendants. I took my purchase to the jewelry store and asked why he didn't sell cubic zirconia. He demonstrated why by placing the cubic zirconia pendant among his diamond pendants and asking if I could see any difference. I couldn't! The jeweler's response was "Now you know why I don't sell them."

I frequently relate this experience when lecturing to high school and college students and suggest that the guys in the audience should never give a diamond. They should go to a department store and buy a cubic zirconia and then go to a jewelry store and buy a gift box for the ring. The gal would never know that the stone in the ring wasn't a diamond unless she had the ring appraised, and "You wouldn't want to marry a woman who didn't trust you." Once when I told this story, a young man stood up and exclaimed, "Sir, I'm not going to give a diamond to a woman to impress her. I'm going to give her a diamond because I love her!" That statement demonstrates the importance of love and illustrates the difference between a career designed to impress and a career designed to serve.

One of the jewels that each of us can give to the world is our career. The quality of the jewel that we give will depend on personal characteristics such as determination, dedication, and distinction. How determined are we to assure that our careers and our lives represent the best that we can offer, especially when our best may require personal sacrifice? The passengers on the September 11, 2001, United Airlines Flight 93 demonstrated determination. With their lives in jeopardy, they struggled to assure that the plane would not be used as the terrorists had planned. When we encounter something that is wrong such as the beginnings of a failure, are we determined to correct the situation? And even if we are determined, do we have the dedication to continue in the face of personal risk? The public safety workers in New York have demonstrated the type of dedication required for giving diamonds. The majority of us will never be called to place our lives on the line; thus, our personal risks will generally be more emotional than physical. True determination and dedication to correctness may require the risk of popularity, position, and creature comforts. Standing up for what is right is difficult when you are standing alone. Have we dedicated ourselves and our careers to excellence when mediocrity is the call of the day? What level of performance is personally acceptable? Do we compare ourselves to the best, the average, or the mediocre?

The events of September 11, 2001, demonstrated to the world that a segment of our population knows how to give diamonds to the public. We have numerous opportunities to serve the world through our careers. Will we have the determination and dedication required to serve with distinction, or will we give the public cubic zirconia, or maybe even cut glass, because we want to impress and really don't love anyone but ourselves?

Practical Failure Analysis, 2001, Vol 1(6), p 4–5

LOOKING BEYOND OUR COMFORT ZONE

Several months ago, my wife and I had the privilege of babysitting two of our grandchildren. One of the children, Ty, was only about 18 months old, and this was his first extended stay under our care. Our other grandchildren, including Ty's big sister Emily, call me Big Mac and call my wife Mamaw. Ty and Emily have another set of grandparents that live very near their home, and consequently, Ty spends considerably more time with those grandparents than he does with us. Emily and Ty's other grandparents are called Mu Mu and Pa Pa. Mamaw and Big Mac don't look anything like Mu Mu and Pa Pa except that we are all old, wrinkled, and don't have to ask to receive our senior citizen discounts. However, while he was under our care, Ty would frequently refer to me as Pa Pa and to my wife as Mu Mu. He was simply living in his comfort zone. Two months later, when Ty's little sister Annie Grace was born, he called us Mamaw and Big Mac. His comfort zone had grown.

This issue marks the first birthday of *Practical Failure Analysis*. The publication is therefore in its infancy and is seeking to establish its comfort zone. Volume 1 contains a reasonably wide variety of case histories and discusses several approaches to the analysis and prevention of failure. The periodical may have reached the goal of "being of value to the failure analysis community." However, before it fulfills my vision of becoming the resource of choice for failure analysts, metallurgists, ceramists, polymer and materials engineers, and anyone wanting to understand component failure and failure prevention methodologies, we must learn to reach beyond our current comfort zone. Frequently, many failure analysts choose to refer to only one set of "grandparents." The analysts that grew up near engineering science and mechanics may call for an analytical approach to a failure problem, while those that grew up near metallurgy may address the failure by examining the microstructure and fracture morphology of a failed component. Other analysts who grew up in other regions may prefer measuring mechanical properties and/or material chemistry. Reference to the articles in Volume 1 suggests that very few analysts focus on the entire tool bag. We tend to be like Ty and simply call for the things we are accustomed to using.

Several decades ago, Rich McNitt, a member of the *Practical Failure Analysis* editorial board, wrote, "Interdisciplinary approaches to design considerations are required if we are to develop techniques for efficient materials utilization." He suggested that the ability to predictably use materials in aggressive environments depends on integration of the knowledge of environmental effects on materials structure-property relationships with an analytical ability to model material response to microstructural evolution. In his appeal for integration of the engineering science and mechanics communities and the materials (metals, ceramics, polymers, and composites) communities, he emphasized that "present and future national needs will require interdisciplinary efforts and cooperative research and development." That appeal was made almost 25 years ago, but many of us still have very confined comfort zones and are avoiding the integrative processes. The integration Rich suggested is clearly slow in coming. For example, recently published articles do not mention the use of finite-element analysis as a tool in failure investigation and rely almost completely on the use of microstructural analysis. This contrasts with other articles which do not show a single micrograph but include several analytical evaluations. These two contrasting styles demonstrate that highly qualified people (and their reviewers) may miss opportunities to expand their comfort zones.

The ultimate success of any failure analysis depends on several factors:

- The quality of the experimental and analytical approaches used to define the failure process
- A demonstration of the root cause of the failure
- Communication of the failure analysis results

- The effective communication of the results frequently requires discussions with people from a large cross section of the engineering disciplines.

To convince those responsible for the disposition/correction of the failure, we must have at least considered the dispositioner's "favorite" failure analysis approach. I know that I appreciate microstructures and fractographs, while Rich McNitt appreciates analytical descriptions, and others appreciate property measurements, chemical analyses, and/or historical evaluations.

Twice a year, I teach the ASM International short course "Metallurgy for the Non-Metallurgist." The participants in this class come from a variety of backgrounds and engineering and science disciplines. I've often noticed that during the first day of class, most of the students are reluctant to discuss their personal experiences, even when those experiences relate directly to the class discussions. This reluctance changes with time, and by the last day of class, virtually everyone becomes involved in the discussions. The participation of the students greatly increases the value of the class, primarily because such participation effectively forces the discussions to become interdisciplinary.

Practical Failure Analysis is an infant and, like Ty, is searching for a comfort zone. Let's force this periodical to reach beyond its current level of comfort and call out to another set of grandparents, and develop into the resource of choice for everyone in the failure analysis and prevention community. Such growth requires the participation of each of you and your willingness to reach out to a multidiscipined world.

Practical Failure Analysis, 2002, Vol 2(1), p 4–5

INTERVIEW WITH AN EXPERT

The ASM International annual event is an assembly of experts: experts in physical metallurgy, materials engineering, environmental degradation of engineering materials, education, training, research, management, and even failure analysis. My wife Fran accompanied me to the annual event in November 2001 and mingled daily with new friends and old acquaintances. We had lunches and dinners with internationally recognized metallurgists and had an evening gathering that included several well-published experts in failure analysis. Fran even accompanied me to a meeting with the faculty and staff of ASM International's Materials Camp. On our way home from the event, I asked Fran, "Who is the best failure analyst you know?" Without hesitation she said, "Johnny!"

Mr. Johnny Moseley is with Tyler's Tire & Auto Center in Aiken, S.C., and has managed the service of our vehicles for the past 15 years. The fact that our only car has over 240,000 miles and our pickup truck has over 130,000 miles speaks for the quality of the service provided. Fran likes dealing with Johnny for several reasons. First, he is an expert. Many times he has diagnosed a problem simply by listening to Fran describe symptoms that she recognizes—noises, quivers, and hesitations. Second, Johnny has established boundaries to his expertise, and when a problem exceeds those boundaries, he suggests that someone else solve it. If asked, he will even suggest the proper expert to solve the problem. Third, Johnny does excellent work and accepts the responsibility for the services he provides. He admits when he makes a mistake and acts promptly to rectify the problem. Clearly, Johnny may be the best failure analyst Fran knows even though her husband has been involved with metallurgical failure analysis throughout our 40+ years of marriage.

My interview with Johnny got off to a shaky start when he stated that his fee would include a no-limit credit card and authorization to travel throughout the world. Fortunately, he ultimately settled for several copies of the published interview.

Johnny is convinced that tire and belt failures are major causes for cars being stranded on the highway. Tire failure is generally associated with improper inflation, failure to rotate the tires, and/or poor front or rear alignment. Uneven tread wear provides an obvious signal that tire degradation is occurring and that maintenance is required. However, these signals are generally ignored. Johnny stated that drivers fall into one of two categories: drivers that pay attention to car maintenance and drivers that basically ignore maintenance. Evidently there are few who fall in the intermediate zone. The drivers that pay attention to maintenance would recognize the signs provided by uneven tread wear, but because of maintenance schedules, they seldom experience tire problems. Drivers that ignore maintenance also ignore the wear patterns that signal the need for maintenance. Johnny's experiences suggest that proper tire maintenance and proper automobile maintenance go hand in hand. Most drivers either have their automobile serviced regularly or don't think of service until a problem is evident.

ASM International recognizes improper service as one of the six fundamental causes of failure. The cost of improper service to the U.S. economy is tens of billions of dollars per year. Most of this cost could be avoided by simply doing the things that we know need to be done. Johnny pointed out that if a car has a timing belt, the service manual for the car will suggest that the belt be changed every 60,000 miles. Failure to change the timing belt can readily convert a hundred-dollar job into a multi-thousand-dollar experience. There is an old television commercial for oil filters that states, "You can pay me now or you can pay me later," suggesting that the cost of an oil filter now is insignificant relative to the cost of an engine repair later.

Johnny is convinced that with regular service, the cars of today will last for many, many miles. He commented on the facts that a car can now go over 50,000 miles without a tune-up and there are some people that never change the spark plugs in their automobile. The lack of need for regular maintenance in one area may cause a driver to ignore the needed maintenance in another area. How often do most of us check the oil, change the belts, and inflate the tires in the vehicles that carry our careers? Do we have obvious signs of uneven tread wear that we ignore until someone else calls them to our attention? How long has it been since you last looked at the maintenance manual for your career and attempted to provide the recommended services?

Johnny assured me that a cheap tire would go over 40,000 miles without a problem if the tire were maintained properly. He also assured me that a top-of-the-line tire might fail with less than 10,000 miles of service if the tire were underinflated. Education, dedication, concentration, and communication are four of the tires that carry our careers. Are we assuring that these tires are inflated and rotated and that our lives are aligned so that our careers can provide the maximum mileage for ourselves and for society? If not, why don't you take a few minutes, look in a mirror, and interview the expert staring back at you.

Practical Failure Analysis, 2002, Vol 2(2), p 4–5

EVALUATING THE SUCCESSES

Coaches spend a lot of time conducting failure analyses. Each time a game is played, there is a winner and a loser. Some would equate the wins and losses with successes and failures, but in any event, both the winning and losing coaches will evaluate all aspects of the game: the pre-game practices, the players' performance, the on-floor adjustments, the attitudes and efforts, and the expectations. A careful evaluation of these parameters will demonstrate that not all wins are successes and not all losses are failures.

A high school coach once described my son Keith's effort in a losing ball game as "the best overall performance he had ever seen from a high school player." The coach was in the last year of a 30+ year career and didn't hand out compliments often. The game was the first round of the district playoffs and, because the team lost the game, was also the last game of Keith's high school career. Days later I relayed the compliment to Keith, who simply said, "We lost, Dad! Don't you understand? We lost!" Realistically, Keith's team didn't have a chance to win, but even days after the game, his expectations remained too high.

Too often we get so focused on wins and losses that we forget to consider how well we played the game. A coach's analysis of a win will almost always include a list of things that should be improved. Similarly, the analysis of a loss will generally include glimmers of good performances.

Years ago, my older son Rick, then a high school junior and starting forward on the basketball team, was playing Keith, then a seventh grader, in backyard basketball. We had lights on the basketball court, and neighbors would frequently use a ballgame as an excuse to socialize. The games generally included numerous neighborhood teenagers, but just Rick and Keith were playing that evening. The games were always to ten baskets, and when Rick and

Keith played one-on-one, Rick always won. But on this night, Keith was ahead, 9–4. Several of us thought that Keith would finally beat his big brother, but the score went to 9–9 and then Rick made the winning bucket. Keith threw the basketball against the wall, hard, then turned to walk into the yard and away from the lights.

A neighbor, Terry, put his arm over Keith's shoulder and said, "Keith, if I were you, I'd be proud that I gave Rick such a good game. After all, he is a lot older and considerably taller than you are. I'd be happy that I'd come so close." I'll never forget Keith's response: "That's where you and I are different, Mr. Weishauer. I'm not happy when I lose to anyone in anything!"

As Keith's father, I had to admire his competitive attitude, but I felt that I missed something in preparing his game plan of life's expectations. None of us can win all the time, and because we can't always win, we need to develop strategies to learn from our losses. Failure analyses are made. Lessons learned are developed and published. We even have evaluations of near misses (things that could have caused an accident or a failure) to explain what was learned from them. However, most of us do win occasionally. We need to take a few lessons from the coaching fraternity and learn to evaluate our successes as well as our failures.

Evaluating a success doesn't mean patting ourselves on the back and saying, "Good job!" Successes provide as many chances for analysis as failures. Unfortunately, many elements of society simply ignore success analysis. Many of us are generally satisfied with a success: "We're on time and on budget. What a success! There is nothing to analyze here. Everyone should be satisfied." Contrast this attitude with the attitude of most coaches, especially successful coaches. They are seldom satisfied.

I played enough organized sports to learn that a victory is not necessarily a success. My coaches could always find ways we could have improved, things that we should have done, and areas where we just barely got by. I've walked into many locker rooms with a smile that was transformed to a frown after the coach's "success analysis" was completed. I once heard a teammate describe one of my coach's expectations by saying, "Coach Bradley doesn't expect perfection all the time; he only expects it when you are playing."

Keith expected to win, regardless of the odds. What are our expectations for the planning, personal effort, and performance associated with our work? Do we want to change anything after we've had a success? How often have you heard, "If it ain't broke, don't fix it"? Most of us have many more successes than failures. We are quick to begin a failure analysis, but aren't there lessons to be learned from each of the successes? What did we do correctly? How could we have improved? What changes should we have made? Did we use the best materials, fabrication methods, and joining techniques? Were we too conservative with our design? Success analysis should offer more benefits to society than failure analysis! Let's start looking at our successes not simply as wins, but as opportunities for improvement.

Practical Failure Analysis, 2002, Vol 2(4), p 3–4

SETTING YOUR GOALS

Watching a group of eight-year-old boys play "organized" basketball can be a thoroughly enjoyable experience, especially when one of the players is your grandson. Austin was part of an Upward Basketball program in Fayetteville, Ga., this past winter. The eight-and-under league was a development league that didn't keep any statistical information about the players. In fact, they didn't keep score. The purpose of the league was for the players to learn the game. Learning seemed to mean at least four things: basketball fundamentals, the team concept, competition, and fun. However, as I watched—totally unbiased, of course—I soon realized that Austin was the best player on the court. The parents and grandparents of other players may have disagreed, but in my eyes, he was the best passer and rebounder, the best shot and ball handler, and the best defender. Austin had had a lot of practice because his dad had been a high school basketball coach when he was younger, but I'm convinced that some, if not most, of Austin's basketball skills were the result of goal setting.

The summer before Austin was three years old, his goal was to make a basket. He would shoot, shoot, and shoot again, seldom hitting the rim and never coming close to making a basket. The basketball goal he used was set at the standard height, ten feet off the ground. The ball he used was the standard high school boys' basketball. The combination of goal height and ball weight simply exceeded Austin's strength. He couldn't get the ball over the rim. One of my treasured photographs from that summer is of Austin trying to make a basket. His form is excellent. He looks as if he has mastered the shot, but he was never satisfied with simply looking good. Whenever he was asked, "Austin, did you make that shot?" his response was always the same: "Not yet!" He never said no. He never asked for someone to lower the goal. He never asked for a smaller ball. He simply shot and shot, working to achieve a dream, working to make the goal. He would accept criticism, incorporate change, and improve his effort. He always wanted to play and would ask anyone who looked like a basketball player to join in the practice. By the time he was six, he could, on occasion, make ten straight foul shots while shooting a standard ball to a standard goal.

The ultimate success of a failure analysis generally depends on two things: the goals that the analyst and funding agency jointly establish prior to initiating the failure investigation, and the analyst's dedication to reaching the goals. When the goals and the dedication are high, the results will be worth sharing with the failure analysis community and generally worthy of publishing. Lowering the goals and/or decreasing the dedication generally result in decreasing the value of the investigation. Practice is the process of preparing, submitting, reviewing, revising, and publishing manuscripts. Some good shots have been made, but the failure analysis community needs an increased willingness to strive toward the highest practical goals and an increased willingness to share.

Goals are relatively easy to set. However, setting an appropriate goal is often difficult. If the goal is too high, the path to the goal may be too costly, too difficult, or take too much time. If the goal is too low, meeting the goal has limited value and the drive toward the goal does not demand our best. Excellence is achieved when the goal is to prepare and publish a high-quality manuscript. Whenever the goal is quality, the manuscript contains data analyses, interpretations, and/or images that provide immediate contributions to the failure analysis community.

Excellence will be fully achieved when the failure analysis community accepts only manuscripts of the highest quality. As is the case with contributions from any technical community, manuscripts submitted vary from marginal to extraordinary. This variation is related to the authors' experiences as well as their goals. Inexperienced authors generally need the coaching that the review process is designed to provide. The manuscripts submitted to a journal are the shots that are taken by the failure analysis community. Some hit the goal of excellence, and some don't quite reach that goal. Review, revision, and editing improve some shots, but some shots will simply fall short. However, the ability to shoot improves with practice if the goal is kept high.

Many of you have contributed to the success of failure analysis journals. Thanks! Improvement and continued success in our journals depend on you and your involvement and willingness to ask others to play in the publication game. The journals need for the failure analysis community to take more shots. We need you to shoot and shoot again, always aiming for the goal of excellence and always willing to consider the recommendations of the review team and editors. Is every published article excellent? Not yet! However, if we keep shooting the big ball at the high goal, excellence will be achieved.

Practical Failure Analysis, 2003, Vol 3(4), p 3–4

FACING THE HARD PROBLEMS

Several years ago I temporarily quit drinking coffee at work, because I noted that whenever I faced a hard problem, I headed to the coffee pot. My excuse was that the coffee diversion enhanced my focus on the problem at hand, but reality was totally different. The coffee was a distraction that allowed me to turn from the difficult and begin working the easy tasks on the agenda. After all, the desk was always overfilled, and the apparent progress provided by completion of the easy problems was enjoyable, allowing postponement of the effort required to face the hard problem. However, the cessation from coffee did not solve the tendency to postpone the difficult. Cola replaced coffee, walking the halls replaced cola, and now coffee has replaced walking the halls. A diversion from the difficult is always easy to find.

Arnold Saul is a good friend who, a few years back, provided an excellent illustration on how to face a difficult task. Arnold was turning 60 and needed a challenge to prove that, even at 60, some portion of the physical ability he had had as a youth was retained. He decided to demonstrate this retention by taking a 60-mile bike ride—one mile for each year of his life. He then told friends, family, and other acquaintances that he was going to ride 60 miles on the New River Trail on April 13 and invited each of us to accompany him. Rain began falling on April 12, continuing through the night and into the early morning. No provisions had been made for bad weather, and the ride began in the rain. In spite of the weather, Arnold faced the hard problems and completed the task.

The New River Trail, which is less than 60 miles long, is an old railroad bed that is relatively flat, but it does have some hills. The limited trail length and the hills make it impossible to coast all the way, but with careful planning one could minimize the uphill portions, focusing on coasting and downhill travel. The coasting approach, however, was in contrast to Arnold's plan, where downhill travel was minimized. He didn't want to coast to completion; he wanted to ride to success. When the ride was over, Arnold returned home to a surprise birthday gathering. The success of his birthday ride hinged on both his physical ability and his willingness to use that ability in the face of a hard problem. Doesn't the success of most of our failure analyses hinge on our ability and our willingness to use that ability while conducting the analysis, especially the hard portions of the analysis?

Arnold's successful ride demonstrates four steps that will help as we face a failure analysis, or any task, that is forcing us to do the difficult: (1) set a goal; (2) inform others that the goal is set; (3) involve others in accomplishing the goal; and (4) celebrate success when the goal is achieved. When facing a difficult task, it is easier to postpone action if no one but you knows that the task must be accomplished. The next time a difficult task is dragging you toward the coffee pot, a cola, or the hall, try telling your secretary, your peer, your boss, or even your spouse, "Today I'm going to accomplish..." The simple act of telling someone will clearly define your expectations. An established goal, coupled with the awareness that others know of the goal, is frequently a strong motivator. Combine this motivator with a request for assistance, oversight, and/or review, and the drivers necessary to face a hard problem should be in place. Clearly, success will hinge on ability, but that ability will have been magnified through motivation and increased involvement. Isn't it amazing that we are often willing to use the inanimate tools in the failure analysis toolbox but are unwilling to tap the animate resources at our fingertips? There is an old country western song that states, "I had your love on the tip of my fingers but let it slip right through my hand." How frequently do we have success at the tip of our fingers but postpone that success because we are unwilling to set a goal and/or ask for assistance in achieving that goal?

The celebration of a technical achievement is also something that many of us neglect. We don't want to brag, so professional achievements often go uncelebrated. Several weeks ago my granddaughter Kate played in her first soccer game. After that game she was celebrating her success (basically the success of simply having participated in the game). She phoned her Mamaw (grandmother) and talked for several minutes, concluding with the comment, "I just can't wait until the tournaments!" Celebration generally promotes a positive look toward the future. We shouldn't be childlike, but our future tasks could benefit through celebration of our present successes. However, the success should be worthy of celebration because it involved our best effort and was not achieved by simply coasting downhill.

The next time you face an extremely difficult task, will you head for the coffee pot, seek out a cola, or will you ride to success with Arnold? Here's hoping that we can all set appropriate goals, involve appropriate teams, and celebrate personal achievements. The achievement might come faster if we learn to face the hard problems and avoid walking in the halls.

Journal of Failure Analysis and Prevention, 2004, Vol 4(3), p 3–4

THE ONLY ONE

Fractography and metallography frequently provide the evidence required for the successful completion of a failure analysis. Macroscopic examination of the fracture surface can generally help establish the direction of crack growth and assist in locating the fracture origin. Microscopic examination of metallographic cross sections can provide indications of the heat treatment, service conditions, and whether the heat treatments were part of the manufacturing process or a result of in-service exposure. Microfractography can establish the mode of failure and identify any changes in fracture mode that accompany crack growth. The positions of fracture mode transitions can be coupled with materials properties to estimate loads, and fatigue striations can be used to estimate cyclic stress intensities at various locations. The analyst, by proper selection and examination, may be able to establish a detailed description of crack growth processes simply by examining selected regions on the fracture surface and determining the microstructure of the material in those locations. The key to success is selecting the right regions to examine, conducting the proper examinations, testing the appropriate pieces, and interpreting the data correctly. To accomplish this you must be...

My wife and I attended a wedding in Richmond, Va., the last weekend in October. The entire weekend was a high-class affair. The rehearsal dinner was in the River Road section of the city, the wedding ceremony was in the chapel at the University of Richmond, and the reception was held at a Country Club of Virginia. I almost purchased new clothes but squeaked by because my wardrobe contained two dark suits, two white shirts, and two striped ties from a different era. My wife Fran, however, used the occasion to justify several purchases and even made me polish her shoes. We were invited because the Nunnallys, Joan and

Butch (mother and father of the groom), have been our friends for decades. Ward (the groom) grew up with our children, as did his sisters Noel and Shannon. Noel and Shannon are both married, have two children each, and had their entire families at the wedding weekend. Joan's brother, Dwight Hash, who has been our friend since we first met in the early '80s, was also at the wedding. We have floated the New River, attended the Old Fiddler's Convention, gone to football games, fished, water skied, dined, and danced with Dwight. We knew that he was a "good ole boy," but this was the first opportunity we had to see him in action in high society.

> The key to success is selecting the right regions to examine, conducting the proper examinations, testing the appropriate pieces, and interpreting the data correctly.

Dwight was remarkable. He helped Shannon care for her infant and toddler, took special interest in Noel's boys, and was at Ward's beck and call. He danced with his nieces, listened to his sister, talked with his brother-in-law, and spent time with the bride's family. Dwight obviously felt that his role at the wedding was to assure that the groom's family was socially engaged and that his nephew, nieces, great nephews, and great niece were happy and content. Without ever taking center stage, Dwight played a key role in making the weekend a thoroughly delightful experience for everyone. As the reception was drawing to an end, I told Dwight that he was setting an awfully high standard for uncles. He glanced at Noel's two boys and replied with a clear statement of accepted responsibility: "I have to be good—I'm the only one they have!"

Dwight's reply is directly applicable to any failure analysis. There is only one correct answer, and that answer is most readily obtained when the analyst accepts the responsibility for "being good." Too often we are given a responsibility to be fast (schedule driven), to be inexpensive (budget driven), and/or to be certain (confidence driven), and the responsibility to be good may be pushed toward the background. Many decisions are based on the preponderance of evidence, but if the analyst isn't good, the evidence collected may point in the wrong direction. The interpretation of every piece of evidence may be correct, but if the wrong pieces are selected, the failure analysis may lead to the wrong conclusion. There may be lots of fracture surface, but there is only one fracture origin. Numerous broken pieces may be collected, but only one piece broke first. A component may contain many inclusions, but only one group of inclusions was at the failure initiation site. And when the analysis is complete, there may be several opinions, but, at best, only one opinion is correct.

Being a good failure analyst is like being a good uncle at a wedding. You will need to pay attention to the little things, dance with expenses, listen to the client, talk with associates, and spend time with a schedule. The key to success, however, is to remember that the real job is to "be good"—because the answer you seek is the correct answer, and to a good analyst the correct answer should be "the only one."

Journal of Failure Analysis and Prevention, 2005, Vol 5(1), p 3–4

INTEGRITY, CHARACTER, AND DISASTER

Noon-time basketball is a common activity providing tension release and exercise for the faculty, staff, and graduate students at many U.S. universities. At Virginia Tech the activity is termed "playing in the NBA," or the Noon-time Basketball Association. Teams are organized after ten or more "NBA members" have gathered, and the team selection process is generally designed to provide a competitive game. The first team to make 15 baskets wins if they are ahead by two or more baskets, and the winning team plays a new team in the next game. If no one is ahead by at least two baskets, the game continues until one team gains a two-basket margin. The losers sit down. Most of the NBA participants hope to play three games before showering and returning to work. Occasionally a game will be extremely competitive, and getting ahead by two baskets becomes very difficult. The importance of winning the game increases as the level of competition becomes greater and the game goes into overtime. During one very competitive game, my team was behind by one basket. The score was something like 27 to 28 and we had the ball. I attempted to throw a pass to Steve Gilmore, but the pass was deflected and went out of bounds. The other team awarded the ball to us until Steve stated, "I touched the ball just before it went out of bounds." Our opponents then took the ball, passed it inbounds, promptly scored, and won the game. A spectator and potential player for the next game said to Steve, "That was a terrible time to be honest." Steve simply smiled and replied, "Integrity is a way of life." This statement prompted some serious discussion, and virtually everyone involved agreed that Steve's assertion was true and had a wide range of application. The focus of the discussion began to center on the value of sports and competition, with most of us agreeing that competition and team sports build character. However, Norm Marriott, one of my long-time friends, disagreed and said, "Competitive sports don't build character, they simply reveal character; character is generally built by spiritual activity."

These two statements have stuck with me for over 20 years: Integrity is a way of life, and our actions in competitive and/or opportunistic situations reveal our character. A January 2005 train disaster near my home in Aiken, S.C., gave me new insight into both of these statements. My understanding of the disaster, its aftermath, and the integrity and character involved, is revealed in the following paragraphs.

A short Norfolk-Southern train was switched onto a side track in Graniteville, S.C., during the afternoon and was left on the siding for unloading the next day. Unfortunately, the switch was never reset to allow a train to pass down the main track. The next train passing through Graniteville barreled onto the side track and crashed into the stationary short train. The crash derailed and severely damaged several cars from the long train. The damage to one of the cars caused the release of chlorine gas to the surrounding area. At lease 9 people were killed, over 200 required hospital treatment, and over 5000 were evacuated from their homes. Businesses, churches, and schools were closed; roads were blocked; decontamination stations were established; and radio and television stations gave advisory updates. The surrounding communities set up emergency contact stations, opened homes and churches to provide lodging for the displaced, and created food kitchens and clothing closets. The first responders, law enforcement groups, emergency medical personnel, hospital staff, Red Cross workers, and numerous other groups and individuals demonstrated that integrity was a way of life and revealed their character with selfless acts that focused on improving a dreadful situation. Virtually everyone offered to help and worked for those caught in the disaster without any thought of personal gain. Sometimes the helper was even placed at risk. The community demonstrated that most people want to obey the Boy Scout Law and be helpful, friendly, courteous, kind, obedient, and so forth. However, a small element of the population saw the train wreck not as an unfortunate circumstance, not as a disaster, but as an opportunity for personal and/or corporate gain.

The evacuation caused by release of the chlorine gas left many homes unattended, and in an attempt to protect those homes, local law enforcement imposed a curfew: no one was allowed in the evacuated area after dark. Some

Failure analysts are seekers of truth and hopefully will not compromise integrity for personal gain.

individuals thought that the combination of evacuation and curfew provided opportunity for theft, and during the first night of the curfew, several people revealed their character by being arrested for sneaking into the evacuated area and the emply homes.

The Norfolk-Southern railroad set up a help desk where displaced individuals could, with proper identification, obtain a sizable check to assist with the expenses resulting from the displacement. At least eight people were arrested for obtaining these checks after going to the Division of Motor Vehicles and changing their addresses so that their new driver's licenses showed that they lived in the evacuated area. Certainly these actions, under crisis conditions, revealed an undesirable character for this group of people.

The content of character was also revealed for some corporations. Apparently, the billing schedule of an energy company required that meters in the evacuated area be read while the evacuation was in effect. In order to avoid any delay in the billing and collection process, the company "estimated" the meter readings at homes of the displaced persons. Unfortunately, the bills didn't state that the meter readings were estimated, and displaced persons obtained bills that suggested that their meters were actually read on a day that no one was allowed in the area. Additionally, newspaper reports stated that many of the estimated meter readings were on the high side. What kind of feelings did these actions invoke, and what type of corporate character and integrity did the bills suggest? I feel certain that the corporation did not desire to appear greedy, uncaring, and/or self-centered, but isn't that what their actions revealed?

Failure analysts are seekers of the truth and hopefully will not compromise integrity for personal gain. We should always attempt to follow the example set by Steve Gilmore and make the honest, high-integrity call, even when we are the only ones who know we have touched the ball. After all, as Norm Marriott stated over 20 years ago, it is our response to tough situations like participation in competitive sport, that reveals the true content of our character.

Journal of Failure Analysis and Prevention, 2005
Vol 5(2), p 3–4

ARE WE THERE YET?

Basketball practice can be both physically and mentally demanding, but when practice is over, the successful player generally hangs around the gym to take a few more shots. He will be tired, have other commitments that must be met, and may even be alone when these after-practice shots are taken. Ten, twenty, or even fifty repetitions are all taken from the same place on the floor, and when that shot is honed (making ten or more in a row) the player moves to a different location and begins to shoot again. Muscles ache, sweat drips to the floor, and fingertips hurt before three or four different shots are honed, but the honing is only temporary and lasts until the next day when the shots are practiced again. Commitment, dedication, practice, and talent merge to bring the player's skills to a level where he believes that his shot has "arrived." In reality, the shot never arrives; it may be nearly perfect, but constant practice is required to make a high percentage of any shot. The mature player never believes that his skills "are there yet." He may know that he is good, but he also knows that constant effort is required to be among the best. The growth toward maturity is not easy, and the best player, as well as anyone else, may have childish moments that relax the personal pursuit of excellence.

One Sunday in December, we were visiting our son Keith and his family in Fayetteville, Ga., to celebrate significant events in the lives of all three of his children: Austin (11), Hunter (7), and Ellison (4). After church we drove from Fayetteville to Warm Springs for a buffet luncheon that focused on Hunter's spiritual commitment at church. The 40-mile drive took approximately an hour, and Ellison asked some version of "Are we there yet?" at least five times. The same questions were repeated as we returned to Fayetteville. Hunter wanted to go to Warm Springs because he thought that the restaurant there was "the best." He wanted the

> The mature player never believes that his skills 'are there yet.' He may know that he is good, but he also knows that constant effort is required to be among the best.

best for his family, but his sister did not want to make the effort required to secure the best. Most preschoolers have not attained the patience that maturity brings; thus, while having the desire, they frequently lack the commitment that being the best demands. Hunter is more mature than Ellison and did not have a single complaint during the entire trip. However, his lack of maturity became apparent when we celebrated an early Christmas upon our return to Fayetteville. Hunter had asked for "moon shoes," and because my wife Fran and I are grandparents, moon shoes were among his early Christmas presents. The moon shoes are basically two mini trampolines; one trampoline strapped to each foot. Learning to walk and/or jump on the moon shoes is difficult, and Hunter ceased to practice long before he mastered the shoes. "I'm going to practice tomorrow" was his excuse for quitting, but a week later he hadn't tried again. Children frequently lack the dedication required to accomplish a difficult or unfamiliar task.

Austin, a middle schooler, received a Star Wars light saber game that he rapidly unboxed, attached to the television, and began to play. He never looked at the directions, and although he had never seen this type of game, somehow he managed the connections needed to play. One of the early sessions suggested that the participant practice several light saber moves, but Austin skipped the practice and continued to play, albeit with limited success. Middle schoolers apparently think they already know it all, and therefore practice is unnecessary. Austin wants to win but is more willing to be a winning participant at an easier level than to put in the practice required to successfully participate at a harder, higher level.

There is a commercial on U.S. television which emphasizes that virtually no one wants to be ordinary at anything. We all hope, study, and train to be one of the best. As failure analysts, metallurgists, mechanical engineers, scientists, or whatever our vocation, the desire to be among the best is always there, even if it is tucked in the back of our minds. But desire does not produce results—work does, and our commitment to work generally determines our level of success.

There are at least three levels of success in any vocation: ordinary, above average, and "one of the best." How would you describe yourself? How would your peers describe you? How would your customers describe you? If the answer to any of these three questions is not "one of the best," then you need to "hone your shot." In fact, regardless of our answer to the questions, we all need to hone our shot. We all need to strive for constant improvement by increasing our knowledge, expanding our skills, and deepening our wisdom. For the failure analyst, honing the shot requires the same commitment, dedication, and practice that it does for a basketball player. Because no one wants to be an ordinary failure analyst, we need to practice our shots by reading, studying, discussing, and thinking. We need to grow from criticism, build from mistakes, and reach for challenges. We either practice, knowing that we can never really arrive, or we relax and become like children, avoiding the effort required to achieve excellence but continually asking if we are there yet.

Are we there yet? I hope that most of us don't really believe that we have arrived, but I also hope that most of us have the commitment and dedication to push our analyses toward ever-increasing excellence and away from the ordinary. If we do some day, on some occasion, our commitment, dedication, and practice will merge with our talents to produce a product that is truly one of the best.

Journal of Failure Analysis and Prevention, 2006
Vol 6(2), p 3–4

RECOGNIZING THE BOSS

My wife Fran and I celebrated our granddaughter Annie Grace's birthday on Thanksgiving simply because we couldn't get together on November 20, the day she turned five. One of her presents was an interactive toy: a mechanical dog named Lucky. The dog came in a box that clearly stated "Batteries not included," so another present was a package of six AA batteries. The box contained a service manual, and the operating instructions were printed in technicolor on one side of the box. Kate, a third grader with a "type A" personality, led the assembly and instructed her little brother and two younger cousins on how Lucky should be treated. "Call his name, and when he barks twice, tell him to perform one of the tricks shown on the box. Lucky can sit, shake hands, stand on his head, and lie down. Now y'all be quiet and I'll show you how he works." After several minutes of clear instructions and wonderful performances by Lucky, Kate turned the dog over to Annie Grace. Unfortunately, Tripp and Ty both thought that Kate had turned Lucky over to them. The result was total confusion. All three children gave commands and Lucky had no idea who was the boss. He sat when Annie Grace thought he should stand on his head; he laid down when he should have sat. Because Annie Grace knew she was in charge—after all, the toy was hers—she simply grabbed Lucky and forced his legs, tail, head, and body into the desired position. Because he couldn't recognize the boss, Lucky became a failure.

Annie Grace, Ty, and Tripp all participated in the failure analysis. Kate even offered the suggestion, "You probably broke Lucky when you moved his legs." Each child would demand that the others be quiet while they attempted to become a boss that Lucky could recognize. Occasionally Lucky would perform the correct trick, but generally he was a failure. They moved Lucky's legs, repositioned his head, and wagged his tail. They looked at the batteries, they glanced at the instructions on the box, and they blamed each other. Each child had an opinion as to why Lucky failed. However, even after the failure analysis was complete and Lucky was repackaged for a trip to his new home, not a single child mentioned that the failure might be attributed to Lucky's inability to recognize his boss.

As the editor of a technical journal, I've read numerous articles, made comments, provided suggestions, and rejected some manuscripts. During the editing process I've come to realize that there are many bosses in the failure analysis laboratory.

Funding is a boss that most of us readily recognize. I've read the comment "due to lack of funding" all too often. Schedule is another boss that wants to take charge of the failure analysis. "Because of the limited time available" is a well-used phrase. Access to equipment is another boss that governs the steps included in the analysis process. Many of us simply use readily available equipment rather than the best available equipment. Education and training are also bosses. There is a tremendous difference in a metallurgist's approach to the failure of a mechanical system and a mechanical engineer's approach to the same failure: metallography and fractography versus a finite-element analysis, and mechanical testing . Some analysts simply use computer modeling. Failures have been attributed to fatigue solely because of the striations on the fracture surface, while other similar fatigue failures have been diagnosed through analysis without a glance at the microscopic features of the fracture surface. Training, resources, and experience become the bosses and the analyst may stop searching before the analysis is actually complete.

Lucky's failure demonstrates that recognizing the boss is critical to the success of any failure analysis. We are blessed with numerous failure analysis technologies and methodologies. We may use fault tree techniques or simply rely on experience to guide the failure analysis process. We are all limited by resources and schedule, but techniques, resources, and schedule should never become the boss of a failure analysis. The only boss we should recognize is reasonable certainty. Phrases such as "limited funding," "restricted schedule," and/or "available resources" should not be included in failure analysis reports. If we allow reasonable certainty to govern our quest for truth, we will not need to qualify our conclusions and demonstrate that there have been too many bosses in the failure analysis process.

Lucky was a 30-dollar dog, but even at that price he couldn't succeed when he listened to a multitude of voices. He tried to respond to several bosses at once only to find that he couldn't stand on his head and sit at the same time. In a similar fashion, we may not be able to achieve reasonable certainty when we think that cost and schedule are the boss's voice. Oh, we may be like Lucky and get the analysis right sometimes, or even have an Annie Grace that bends the analysis into the correct position, but real success can only be achieved if the failure analysis process leads to a conclusion that provides the customer with reasonable certainty.

We all hope that our failure analysis processes are lucky and that reasonable certainty can be readily achieved. However, when they aren't, let's not allow a multitude of bosses to lead us along a path toward failure.

Journal of Failure Analysis and Prevention, 2007
Vol 7(1), p 1–2

WAITING IN LINE

Traditionally, grandchildren visit us during their winter breaks from school. The visits include a trip to Winterplace to snow tube. Regulations require that a potential snow tuber be over 44 inches tall and capable of riding a tube alone—no help from mom, dad, or grandparents. These regulations challenge our five-year-old grandchildren, but the youngsters come with their immediate family and can depend on older brothers and fluffy hats to help them meet the requirements. One family will come on the Monday of Presidents' Day, while the other family will come on a non-holiday Wednesday. The experiences of the families will be vastly different. Our past experiences suggest that each child in the Monday holiday family will be able to get five or six tube rides during the session, while each child in the Wednesday family be able to go down the slopes over twenty times. Crowds make a tremendous difference in the nature of the experience. The Monday holiday family will wait in long lines for a ride to the top of the slope and the ride down the hill; the time spent in the snow tubes will be minimal, but time spent waiting to ride will be extensive. The reverse will be true for the Wednesday family. The cost for the experience will be the same for each child, irrespective of the time spent waiting to ride.

The concept of waiting to ride has become so acceptable that there are consultants who guide the experience in many offices and recreational facilities. The medical profession moves patients from one waiting room to another simply to create the impression of progress, while the lines at Disney and other recreational parks are constructed to hide the real lengths and may simply move customers from a line into a waiting room. There are three things I find unacceptable in most waiting-to-ride experiences: waiting past the appointment time, line cutting by other customers, and the lack of "corporate" concern about the first two. Improper service is one of the six fundamental causes of failure, and, unfortunately, modern society is rapidly becoming saturated with corporations that have virtually no concern for the waiting in line failures.

With answering machines, smart phones, e-mail, fax machines, card readers, and creative line management, talking to a real person is difficult. The airlines now use machines to check travelers onto flights. Grocery stores have rapid checkout lines where the customer may select, scan, weigh, and package their own groceries without personal contact with anyone in the store. Service stations have become gas stations, and customers now pump their own gas and pay at the pump with a credit card. Credit card companies often charge, or attempt to charge, their customers fees for the privilege of possessing a "silver," "gold," or "platinum" card. Certain restaurants refuse to provide separate checks for large parties. One can browse in many stores for hours without ever hearing the words, "May I help you?" Some tradesmen say, "I'll see you on Tuesday," but fail to specify which Tuesday they mean.

Service is a disappearing quality in many professions. Military personnel, public safety, fire, and police officers, emergency medical personnel, and school teachers are major exceptions, and most of these people are dedicated to public service. However, the concept that an elected official should be a public servant and statesman has been replaced, and current politicians are too often self-serving in their quest for personal wealth and/or power. They try to build themselves up by tearing down others and build an image based on lies and promises that will never be fulfilled.

The lack of real service and dedicated servants in our society can have an adverse influence on the failure analysis profession. Our profession is a service profession and we need to carefully avoid the current trend away from customer care. How often do you hear the words, "Please hold. Your call is important to us and we appreciate your waiting"? Do your customers hear those words? How often? Soft music in the background doesn't make the waiting any easier and neither does answering the phone with "I'm sorry you had to wait." Service—real service—requires that the server care about the customer. When the average customer waits over ten minutes before the phone is answered by a person, perhaps the server has more concern for the bottom line than for the customers waiting in line. Most customers will be glad to wait on a few occasions, but customers shouldn't have to wait an extensive period every time they call.

As failure analysts, we should recognize that improper service is one of the basic causes of failure and should operate to assure that we always render the proper service. Good service is not only good for the customer, it is good for business. Rendering the proper service requires the server to address the other five fundamental causes of failure. Proper service requires proper planning so that improper design doesn't render the service inadequate. Proper service requires selecting the right phone and messaging system and the right administrative assistance to assure that the proper materials are available to both the server and the customer. Proper service requires that the failure analysis team is trained to respond to customer questions and needs, thus assuring that improper processing of the staff doesn't lead to improper service. Most service plans and a few personnel contain defects that must be recognized and corrected so that defects don't lead to inadequate service. Finally, as the service system is assembled, the big picture of both customer and server requirements must be addressed so that improper assembly doesn't lead to inadequate service. The failure analysis community, because of the nature of the business, should be comprised of leaders in customer service who don't leave their customers "waiting in line."

Last year, two of my children's families went snow tubing during visits to our house. One family tubed on a holiday and ended up spending most of their time waiting in line. That family is not returning for a snow tube adventure this year. The other family went on a normal weekday and spent most of their time tubing, and they are returning for another non-holiday experience. The family going on Presidents' Day this year hasn't been there before, and I'll bet that this will be their last trip to the snow tube slopes. Waiting in line simply isn't good for business, regardless of the images presented by many modern corporations.

Here's hoping that all your customers will spend more time on the slopes and less time waiting in line as they visit, e-mail, and phone your company during 2007.

Journal of Failure Analysis and Prevention, 2007
Vol 7, p 77–78

IS IT JUST A JOB?

Delmer Wyatt, a friend and master carpenter, was building his first speculation house. He personally did most of the construction but occasionally needed to hire additional help because some jobs simply require at least two workers for successful completion. Delmer was part of the crew that did the renovation of our house. He has a wonderful work ethic, takes great pride in his work, and loves being a carpenter. Actually, Delmer apparently loves most of the processes required for house construction. Several weeks ago, he was installing some cornices that he made for our house after my wife Fran found a need for window dressing on three of our picture windows. While he was installing the cornices, we were discussing the progress on his "spec" house, and I felt that the story he told was appropriate for an editorial.

Delmer, looking for someone to help him install the hardwood flooring in the living room, approached another carpenter who was working for a mutual acquaintance and asked, "Is there a time when you could work with me to install some hardwood flooring?" The answer, "Oh, just about any time... after all, this is just a job," surprised him. Perhaps "surprised him" is too weak of a statement. Delmer immediately told the potential helper that if he thought carpentry was just a job, he didn't want the man working on his house. Delmer believed that any good worker felt that his contributions went far beyond being just a job. As Delmer finished installing the cornices, he ask if I thought he was being too rough on the potential helper and inquired, "You never felt that what you were doing was just a job, did you?"

My reflections on Delmer's question caused me to cringe. There have been times when I thought that an assignment or task was just a job. I hated doing safety inspections and treated them as if they were just a job. Staff meetings were also generally "just a job," as were fire drills, practice shelter alerts, most training sessions, and virtually all group meetings with senior management. Each of us probably has a collection of "just a job" tasks that we are required to perform on a routine basis. Unfortunately, the quality of our performance on these tasks demonstrates the cause for Delmer's concern over hiring anyone who thought carpentry was just a job.

When an assignment is just a job, our heart isn't in the performance, and all too frequently, neither is our head. We go through the motions, focused on finishing the task rather than on accomplishing the task. My high school basketball coach was one of my heroes and mentors, and even now, years after his death, he continues to have an influence on my life. During my senior year, we won the Group II Virginia State Basketball Championship partially because we could shoot foul shots. We practiced foul shot shooting every day, and before we could leave practice, each team member had to make 25 straight. The task was to shoot foul shots, but accomplishing the task required making 25 straight, and a bad shooting day for any individual reflected on the entire team. Everyone—each individual team member—had to successfully accomplish the task before anyone could go home. The last person to make 25 straight had considerable pressure to perform. The pressure made all of us keep both our hearts and our heads focused on accomplishing the task. The pressure of practice helped us perform in a game, and as a team we made over 75% of our foul shots. There are professional basketball teams today that don't shoot 75%. We shot well in games because of the effort we made during practice.

During my career at Savannah River National Laboratory, we had an occasional safety violation, an occasional injury, and an occasional near-miss. How many of those occasions occurred because of my "just a job" performance during a safety inspection? How often did my lack of attention disrupt a staff meeting or training session? How often did I treat something as just a job rather than using the task as an occasion for accomplishment? As failure analysts, we often face rather mundane tasks that most of us would rather avoid. Accurate and complete note keeping, entries into log books, precisely following established and perhaps required practices, placing scale markers on micrographs, documenting conversations, and discussing costs and schedules are tasks that can be just a part of the job but are certainly worthy of our full attention and careful consideration.

Each of us should reflect on the tasks that we are required to perform but, unfortunately, consider the task to be just a job. We should try to establish a pattern of performance that causes us to fully accomplish these tasks rather than just finish them. After all, none of us wants to have Delmer listen to our discussions, read our reports, or see our notebooks and say, "I don't want you working for me because you think that some part of the failure analysis process is just a job."

Journal of Failure Analysis and Prevention, 2007
Vol 7(4), p 227–228

EXCELLENCE, TROPHIES, AND OTHER AWARDS

I once had the privilege of playing in a father-son T-ball game with my grandson Tripp. Tripp's dad, Dave, was out of town on business and had requested that I act as a substitute. The game was designed to acquaint the T-ball players with the structure of baseball: After you hit the ball off the tee, run to first base, then second, then third, and finally home. Stop running only on a base, and try not to make an out. If the other team is batting, try to field your position, and when you catch the ball... These concepts were foreign to many of the players, and after the coach had reviewed both offense and defense, he asked, "Are there any questions?" The immediate response was, "When do we get our trophies?" I knew that five- and six-year-olds have very short attention spans, focus on the snacks rather than the game, and often view T-ball as a chance to play in the dirt—but trophies? The question surprised me. When do we get our trophies? They hadn't played a game, the achievements of the team were minimal (they had practiced), and no exceptional athletic talent was apparent. Trophies?

I was sharing this experience with my son Keith, who used to coach in high school. He was only too aware of expectations of players and parents. If you try out, you should make the team. If you make the team, you should start or at least get to play most of the time. And when the season is over, you should get a trophy, regardless of the team's accomplishments or the skill of the individual. The concept that everyone gets a trophy was new to me. I played basketball in junior high, high school, and college and never received a basketball trophy, although I did receive a small metal when we won the state basketball championship. All eight of my grandchildren have received trophies, many of which were given to them (at their parents' expense) simply because they played on a team, danced in a program, or played a recital. What happened to the concept of rewarding excellence?

My son Rick demonstrated the quest for excellence through two of his children. Ty and Annie Grace are on a swim team that gives ribbons and trophies to virtually everyone; the entire swim league does. There are bronze swimmers, silver swimmers, and gold swimmers in each age group. The slow swimmers are in the bronze group, the silver swimmers are a little faster, and the best in the age group are gold swimmers. We went to the league championship swim meet and there were at least 18 ribbons awarded in every age group: 6 ribbons for each category of swimmers. I asked Rick, "How can you determine excellence when a bronze swimmer gets a blue first-place ribbon for swimming slower than a gold swimmer who only receives a sixth place ribbon?" Rick responded, "Ty and Annie Grace are not trying to win ribbons, they are trying to swim their personal best times. Excellence is doing your personal best." As I pondered Rick's response, I recognized that even though the times have changed and I'm an old fogey for wanting to limit trophies, the quest for excellence has never changed. Excellence has always been measured not by what you achieve, but by how close your achievement comes to representing your personal best.

As failure analysts, we are often involved in a mess that was created because someone else failed in their quest for excellence: short cuts, design compromises, postponed maintenance, and neglected quality. The root cause of numerous failures is the lack of a quest for excellence among designers, manufacturers, and material suppliers. However, as we analyze someone else's failures, let's not be willing to submit to the same conditions that caused the failure in the first place. Failure analysis trophies are not small shiny brass figures attached to wooden or plastic bases; they frequently come in the form of spreadsheets, cost, and schedule. Often our customers try to award these trophies before we have even begun our analysis, and, like T-ball players, we want to accept the trophies before the task is completed. The analysis will barely be underway when the trophies start being presented: "There is nothing obviously wrong with the material, is there?" "The fracture looks like fatigue, doesn't it?" "Speedy Joe's Fracture and Odd Job Company concluded that stress-corrosion cracking caused the failure. Can you prove otherwise?" By answering yes to such questions we accept trophies for participation without engaging in a quest for excellence. How can we give our personal "best" when we have already given our personal "guess" and now, subconsciously, will try to find evidence to support that guess?

The failure analysis business is populated by a wide variety of laboratories—laboratories that have multi-million dollar investments in analytical equipment, laboratories that focus in metallography and scanning electron microscopy, laboratories that have agreements with universities and corporate research centers, and laboratories that have one man with a magnifying glass. Failure analysis almost always requires teamwork, and, at least in my experience, there are numerous areas where other team members are the better experts. Sometimes my personal best is simply recognizing that someone else could do a better job. I may have tried out for the team but excellence requires that I step aside. Recognizing the skills and limitations of our laboratories and of ourselves is a necessary step toward excellence. Some laboratories are like bronze swimmers, some like silver, and some like gold, and all that any customer can ask is that the laboratory personnel give their personal best throughout the analysis. Regardless of which type of laboratory we represent, we need to recognize that we aren't searching for trophies, we are searching for the truth, and premature, undeserved trophies and awards can place us in a ball game before we are ready to play. The quest for excellence is an ongoing process that we can achieve only by providing our personal best, and to offer any less will make us like the player who wants a trophy simply because he participated in a practice one time.

Journal of Failure Analysis and Prevention, 2007
Vol 7(5), p 303–304

PHYSICIANS AND THE FAILURE ANALYSIS PROCESS

On Tuesday, October 2, 2007, my wife Fran blacked out while driving in Radford, Va. The "driverless" car drifted across traffic, collided with an oncoming vehicle, and finally came to rest on the wrong side of the road. No one was seriously hurt, although the driver of the other car received a few stitches in her head. Fran was transported to the hospital and immediately began to undergo a series of tests designed to find out why she had blacked out. The emergency medical personnel drew blood while she was en route to the hospital. Soon after she arrived at the hospital, Fran was given an EKG, a CAT scan, and a series of X-rays and was interviewed by an emergency room doctor. The emergency room doctor contacted our family doctor and appointments were established with a cardiologist, a neurologist, and our family doctor. On Wednesday Fran had another EEG and was fitted with a 24-hour Holter monitor. On Thursday she was refitted with another 24-hour Holter monitor, had another EKG, and met with our family physician. On Friday the Holter monitor was removed and she met with the neurologist. Today—Monday, October 8—she is awaiting an appointment with the cardiologist, and on Wednesday she will have a glucose tolerance test. We have heard mention of an epileptic seizure, cardiac arrhythmia, diabetes, and several other potential causes. One of Fran's EKGs was "borderline," the other "abnormal"; she had excess electrical activity in one lobe of her brain; her blood pressure was too high, her pulse too fast, and... We have been overwhelmed with information, but, in spite of abundance of data, the cause of the blackout remains unknown. They even have a special term for her condition: Fran had an "unexplained syncope."

This is the first time I have truly realized the similarity between the practice of medicine and the failure analysis and prevention process. The medical profession is conducting a failure analysis in an attempt to explain Fran's syncope. Most of our doctor visits were simply trips to determine the cause of a failure. Why did my wife have this syncope? For the first time in a long time, I am the customer ordering the failure analysis, and I am afraid that I am an awfully poor customer. I want the doctor to tell us what he knows, explain the lines on the EKG, and report to us as soon as new data is collected. It is very difficult for us to simply sit back and wait. We want speculation: Give us a hint about what is wrong. What is your best guess? Should we begin medication to prevent another seizure? We wait for doctors' calls, expect immediate responses, and really do not want to give them time to reflect. We push for their best guesses. Unfortunately, the legal profession helps us force the doctors to guess. We have a right to the test data, even though we do not understand it. We have a right to hear what the doctor is thinking, even though he does not have all the data he wants or needs. Therefore we hear about—or, because we have access to the records, read about—epileptic seizures, cardiac arrhythmia, and diabetes. Much of what we hear and read is speculation, and, as a failure analyst, I should know better than to ask for speculation. I should be patient. Perhaps the need for patience is why those who visit the doctor are termed *patients*.

Reflection on the similarity between a physician's failure analysis and a typical failure analysis report brought several things to light. First, there is the importance of records. When we picked up Fran's medical records from the emergency room, I was amazed at the number of pages. There were solid data: blood work, X-rays, EKG plots. There were less rigorous data such as doctors' notes and interview records. Some of the data confuses us: for example, one report described having SOB, which concerned us until we found it stood for "shortness of breath." Second, the doctors truly want to find the answer. They are curious, interested, and are pushing to find out why Fran blacked out. Except for the family physician, whom we have known for years and was once a backyard neighbor, the doctors demonstrate personal concern without becoming overly involved with our problem. Given the choice, they will not speculate, especially about solutions. The neurologist even said, "You have asked me the same question five different ways and the answer isn't going to change: I can't give you any assurances against another syncope, regardless of what we do. We need all the data to give a proper diagnosis." Third, the doctors are willing to consult with one another. We know of at least six phone calls among the various physicians. Each doctor is comfortable within his area of expertise and feels no shame in saying that he needs to consult an expert to help address a specific concern.

Records, concern, and consultation are just a few of the similarities. I am convinced that the failure analysis community could learn a lot by watching physicians practice their trades. Furthermore, I am also convinced that physicians could improve their practices by examining the failure analysis process. As failure analysts, we will hold several conferences within the next year. If you agree that there are similarities between our profession and the practice of medicine, please ask a physician friend to consider participating in a conference. With several physicians speaking to our assembly, perhaps we can learn from each other. After all, both groups do investigate failures.

Journal of Failure Analysis and Prevention, 2007
Vol 7(6), p 379–380

BULLYING

My wife Fran and I were having a school lunch with our eight-year-old granddaughter, Annie Grace, when we noticed the sign on the bulletin board: No bullying! The sign had probably been prepared by a third grader and depicted a rather large boy standing over a much smaller child. The facial expressions showed that the large boy was forcing his will on the smaller child. No bullying! The concept sounds appropriate at first glance, but any child growing up today is taught to bully almost from birth. All we had to do was look at a larger section of the bulletin board to see an example of bullying being promoted by the school.

The bulletin board was located where any parent or grandparent having lunch with a child would notice the display. The display was also adjacent to the visitors' table where visiting parents eat lunch with their children. The title of one display on the bulletin board was "Star Families." Under the title was a collection of stars descending across the board, with each individual star fully in view. Care had been taken to assure no overlap so that the names on the stars were easily seen. One star was named The Louthan Family. This identification showed that Annie Grace's parents had donated a certain minimum amount of money to the school. The family name of every child in the school wasn't on the board. In fact, it is a reasonable guess that less than half the family names were listed. Why were the family names displayed on a bulletin board located where virtually every child and every lunchtime visitor saw a list of contributors and, by inference, a list of those who hadn't contributed? The children want to see their family star on the board. When only the contributors are given stars, isn't the school bullying the parents into a contribution? The school, just like the large boy, would deny any intent at bullying, but what would the facial expression of a child show when they searched for their family star that wasn't present?

Last evening, a television commercial showed two rather old, obese, ugly men bullying a hotel clerk into reducing the room rate. On being successful in their bullying, the perpetrators proudly stated that they negotiated great rates. No bullying? A football player stands over a tackled and injured player, snarling and celebrating his success. No bullying? A basketball player dunks over a smaller player on an inferior team and then gloats his great achievement. No bullying? Television producers focus favorably on acts of poor sportsmanship, and prime-time shows demonstrate that the strong can defeat the weak and impose their will on the defeated. No bullying? No wonder children pay minimal attention to cries against bullying when examples of bullying permeate their lives. Big trucks bully little trucks, little trucks bully small cars, small cars bully motorcycles, motorcycles bully bicycles, and bicycles bully pedestrians. Finally, perhaps, pedestrians walk home and bully their children, who go to school the next day and see a sign—No bullying!—on the bulletin board.

Bullying is a constant in the media, especially in the commercials. One telephone company bullies with the biggest map while another bullies with the fastest network and yet another bullies with the fanciest phone. "I'm the best" is the cry of the day, but no one defines how the best is measured. Fortunately, many of us can remember when a standard saying was "If you have to tell someone how good you are, you probably aren't very good."

Bullying is common in the political arena, on the athletic field, and even in service and religious organizations. Society seems focused on bullying. Years ago when a group of children was selecting teams for a ball game, the leader generally tried to be sure that the sides were selected to provide competition, not domination. Today domination seems to be the goal of team selection, even when the selection process is simply to provide sides for a game. Excellence is certainly a worthy objective, and the quest for excellence should be a personal ethic for everyone. However, should a road to success be paved by the discarded bodies of people bullied into submission?

One of the fundamental causes of failure is improper service. Service, when everything is properly considered, is simply a measure of how people and, in the case of engineering failures, systems are treated. Success often promotes confidence, self assurance, and the belief that we are correct; and when we are certain of our position we may try to force our will on others. Do we bully our competitors, our customers, or our employees? There is a fine line between bullying and convincing. As failure analysts, we must be certain that we have correctly interpreted the evidence, but we also must be certain that we convince others through education rather than bullying them into submission.

How do your employees, your customers, your suppliers, and your competitors feel as business decisions are made? Are you seen as an old, fat, ugly man bullying them into lowering the room rates, or are you a friend, extending a hand that helps everyone meet their goals?

Journal of Failure Analysis and Prevention, 2010
Vol 10(3), p 167–168

S'mOREOS

We have a family tradition that is celebrated every summer during the Virginia Tech Basketball Camp. At the end of each day's activities, our grandchildren who are attending the camp sit around our kitchen table and talk while eating Oreo cookies and drinking cold milk. They discuss their successes and failures while taking the cookies and dipping them in their milk. As the milk-soaked cookies are eaten, we hear sighs of satisfaction and discussions of the proper dipping techniques. This past summer we had three grandchildren gathered around the table with the double-stuffed cookies my wife Fran—"Mamaw"—had purchased. Hunter, our rising seventh grader, promptly disassembled and then reassembled the stuffing-covered portions from two cookies to form a quadruple-stuffed cookie that he dipped and devoured. A pile of plain, unstuffed chocolate cookies began to grow around Hunter's plate, and Tripp, our rising third grader, asked,

"What are you going to do with those plain cookies?" Ty, the rising fifth grader said,

"You could use them to make s'mores."

We frequently prepare s'mores (a graham cracker sandwich containing part of a chocolate bar and a marshmallow) by roasting the marshmallow over a charcoal fire pit. This is traditionally an evening activity that gathers the families around a fire for fun and fellowship. Thus, the idea of using the cookies to make a campfire treat folded nicely into a typical evening at Mamaw and Big Mac's house.

Tripp, who loves s'mores, immediately said,

"Yeah, and we could call them s'mOreos!"

Hunter and Ty both agreed that "s'mOreos" would be an appropriate name and that we should prepare them as soon as we had a chance. Basketball camp ended the next day, but it was several weeks before we found time for s'mOreos. The Louthan family has now prepared s'mOreos on many occasions and I doubt that we will ever make standard s'mores again. The family totally agrees that the chocolate Oreo cookies make the best s'mOreos and that the stuffing should be included in the treat. Although we tried many other varieties of Oreo cookies, the old standard chocolate certainly provides a superior tasting treat. Many evenings of preparing s'mOreos around the fire with guests of all ages have confirmed the quality of the idea and shown that old traditions can be improved. The creation of a s'mOreo demonstrates three basic principles that can contribute to the success of any failure analysis or analyst. First, improvements can be made in many processes, even when there is a strong tradition and an even stronger urge to continue along a specific path because "we have always done it this way." Second, ideas for improvements often emerge from discussions that are seemingly unrelated to the improvement process. Third, a freely flowing discussion involves observing and listening as well as talking.

Hunter wasn't thinking about anything but eating milk-soaked Oreos with lots of stuffing as he prepared the quadruple-stuffed cookie. Too often we become so focused on the analysis being conducted that emergent opportunities that should be apparent are overlooked. We seek a solution to a specific problem and fail to see applications in other arenas. In fact, the tunnel vision provided by the focus on the analysis may lead us to ignore emerging opportunities. Tripp recognized that Hunter was creating an opportunity by building a pile of unstuffed chocolate cookies and was interested in how that opportunity might be addressed. Ty observed Hunter's pile of unstuffed cookies, listened to Tripp's question, and provided a suggestion. The creation of s'mOreos took all three grandchildren, but it also took a willingness to listen to each other and accept suggestions for improvements to a longstanding tradition. No one planned to cause the movement from s'mores to s'mOreos, but as the discussion flowed and ideas were exchanged, an improvement was made—not an improvement on eating milk-soaked Oreos, but an improvement on a seemingly unrelated process.

As failure analysts, we often disassemble systems and components and look for ways to move from double-stuffed to quadruple-stuffed cookies. In the process, we create a pile of leftover cookies that no one seems to want. As the term *think green* becomes more and more important to both government and industry, we need to find ways to improve traditional processes, constructively use any leftovers, and create new products for our companies and customers. To do this, we need free-flowing discussions, new perspectives, and a willingness to change.

Are you able to change a tradition and move from s'mores to s'mOreos, or are you going to stand on tradition because you have always done it that way—thus missing a potential opportunity for a new process or product?

Journal of Failure Analysis and Prevention, 2010
Vol 10(6), p 435–436

LEARNING LEADERSHIP

A young co-worker was more than a little cynical in describing a "typical" engineering graduate entering the workforce in 2011. He was concerned about their work ethic, their desire to learn, and their organizational skills. Basically, my friend felt that today's youth think that their life should focus on fun and that neither work nor responsibility should interfere with social activity. Many, if not most, students at research universities enter college with at least a semester of "college credits" awarded for high school work. Some students have enough credits to bypass their entire freshman year. These so-called college credits enable the student to take only 12 credit hours each semester and still graduate in four years. If the student studies two hours for each hour in class, this "full-time load" equates to a 36-hour work week with no overtime or homework. Dating, partying, and other social activities couple with a little sleep and overeating to fill the other 152 hours in the week. There is no penalty for skipping classes; attendance is typically not taken, and extracurricular activities often provide an excuse for missing class or having a late homework assignment. The real world bears little resemblance to college life and the shock of a full-time job is often hard to bear. Unfortunately, the typical U.S. education does not adequately prepare its graduates for the real world. Frequently the professors guiding the students have been in academia for their entire careers and have never personally experienced a real-world job. It is no wonder the graduates are not prepared to work!

The next time I saw my friend, he described an at-home experience that caused him even more concern about today's youth. His daughter and a group of friends were playing basketball in the backyard and asked him to establish the teams. Apparently the girls had never chosen sides, and no one wanted to select from the group because not being chosen first might hurt someone's feelings. On reflecting about this experience, my friend noted that his daughter had played organized sports for several years, but he was not certain that she had ever simply gone to the playground to play a "pickup" game. We organize our children's lives to the point that adult leadership is all they ever experience. Mom and Dad, the coach, the band leader, the choir director, and other adults tell them where to go, when to get there, and why they should participate. There is little free time for many middle-class children. School, extracurricular activities, church, weekend retreats, summer camps, cell phones, computer games, and planned vacations fill the days. Parents frequently have to check their children's athletic activity schedules before they can plan the family summer vacation. Full schedules aren't bad, but how do children learn or develop leadership skills when everything they do is organized by adults?

The next generation of failure analysts is being developed, and regardless of the outcome, the new generation will be developed from the current crop of college graduates. This means that the new failure analysts must arise from a population that is frequently accused of lacking organizational and leadership skills and being devoid of the work ethic necessary to excel. The absence of highly skilled college graduates is clearly not universal. Graduate classes also include many skilled, motivated youth that are anxious to learn, willing to work, and desiring to contribute. Unfortunately, in order not to adversely impact the self-esteem of the student, our education system has avoided separating the best from the average and the average from the poor. The standard grades are As and Bs, good references are easy to obtain, and on paper, everyone looks like an ideal employee. The recent motion picture *Waiting for Superman* summarized some of the weaknesses in the education system and demonstrated why the system often drives toward mediocrity. How can a system that refuses to distinguish good teaching from bad teaching and pay the good teachers for excellence ever develop performance-driven adults? The employee—in our case the failure analysis firm—is left to develop leadership in new hires, unless they are lucky or hire someone with military experience. My son Rick said that he learned more organizational and leadership skills in four years as an officer in the U.S. Army than he learned during 16 years of school. He also learned how to make a decision or choice by thinking about the mission without being restricted because of potential self-esteem issues in the people under his command. Life generally evaluates performance and rewards excellence.

An article by Bobby Kiger (*The Foothill Times*) reminded me that it isn't just the youth who are lacking. Mr. Kiger stated that "Young people are not the only ones with (an) 'it's all about me syndrome.' Adults are and can be far worse. The family unit is broken—." I've modified his list of the top ten real-life rules to develop some lessons in leadership:

- Life is not fair, so don't expect to be treated like everyone else. Relish the difficult assignments.

- Real work requires tremendous physical and mental effort, so expect to work overtime and require others to do so.

- The answer isn't generally found in a book but is found in the experiences of the prepared, so plan discussions with those that know the job.

- Preparation requires time, devotion, and study. Make friends among the successful and devote time to study every day.

- Find high-performing mentors and follow their examples, including their involvement in professional and service organizations.

- Experience joy in the work and love the learning. If you don't find joy, you are in the wrong profession.

- Express joy and share happiness with family, friends, and the community.

- Leaders are involved in many things, and work is not the only brick in a successful career.

Teaching leadership to the next generation of workers will require preparing the unprepared. We must teach them that there are winners and losers, that life doesn't start over with each assignment, and simply meeting the minimum requirements does not contribute to success.

I'm certain that the old-timers in my organization thought that my generation of engineers was an ill-prepared group, but they compensated for our shortcomings by requiring that we learn leadership. We need to do the same for a generation whose standards for living have emerged from a society that believes "if it feels good you should do it" and that the good things of life should be given to everyone with no work required. We will accomplish this task if we lead by example and show the world that we love what we do.

Smile and keep on smiling while you do your very best!

Journal of Failure Analysis and Prevention, 2011
Vol 11(3), p 175–176

ATTENTION TO DETAIL

REVIEWING A DO-LOCK

My three-year-old granddaughter, Kate Lilly, loves to draw but has considerable difficulty accepting help from her parents, her Mamaw (grandmother), or anyone else that may be reviewing her artwork. Last week she drew a Do-Lock (Fig. 1) for inclusion in this editorial. Only Kate (Fig. 2) fully understands the Do-Lock, and when questioned, she is more than willing to share her understanding with any reviewer. However, she is basically unwilling to accept any recommendations that require changing the pictorial presentation.

Once a semester I give a lecture to students in the Mechanical Engineering Department of Clemson University. The lecture is part of the ME-400 class, Senior Seminar, taught by Prof. Chris Przirembel. Virtually every time I've lectured, the class has been working on an assignment that requires the students to learn and practice the peer-review process. Each class member prepares a paper and then another student in the class reviews the "manuscript." Prof. Przirembel grades both the manuscript and the peer review.

Engineering careers generally afford numerous opportunities for participation in a review process, and the strength of the career is affected by the engineer's response to the reviewer recommendations. Review and rework are an integral part of most engineering achievements, whereas refusal to accept review and/or criticism provides pavement for the path to failure. My pride has been hurt all too often as a reviewer pointed out weaknesses, inconsistencies, and even technical mistakes in manuscripts I have submitted for publication. My computer even insults me by underlining errors in spelling and grammar while I type. Friends mention mistakes made in oral presentations and management discusses opportunities, commitments, and schedules. Fran, my wife and my grandchildren's Mamaw, evaluates tasks completed and promises made, and Kate will remind me that "we don't say that" if I refer to something as stupid. Review is simply a part of life. Our response to the review process has a major impact on both our happiness and our success.

We participate on both sides of the review process. Conducting an appropriate review is as difficult as accepting the constructive criticism that generally results when a good review is completed. I've had to argue with students, authors, management, and fellow engineers, even when they were clearly wrong, simply because I failed to prepare a sensitive, understandable review. However, the worst thing I've done while conducting a review is to overlook a mistake. The goal of technical review is to improve and validate the quality of technical work. Unfortunately, even with peer reviews, galley proofs, and the associated reads and rereads, errors often accompany the transformation of a manuscript into a published paper.

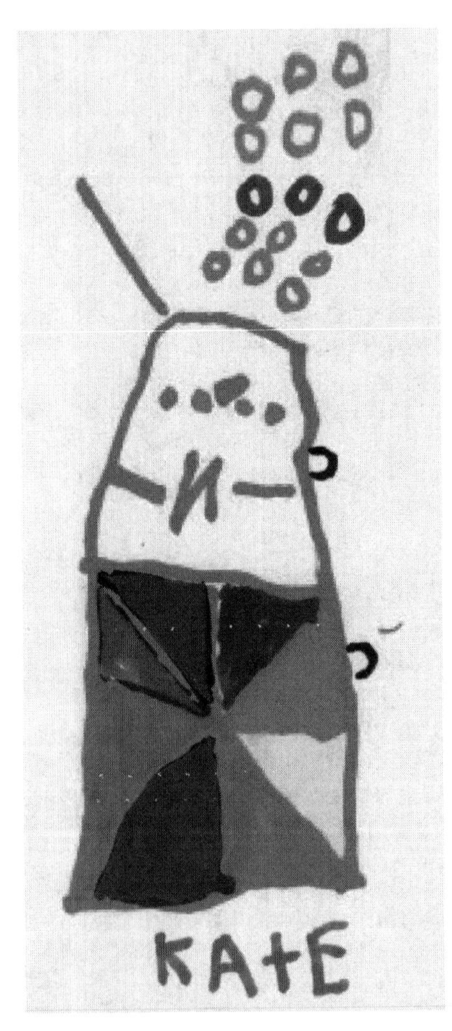

Fig. 1 The Do-Lock is a machine that makes fruit stars. There is a lid, buttons that make the Do-Lock work, and holes where the fruit stars come out. The lid is open. Strawberries, blueberries, and oranges are falling into the Do-Lock. The bottom is decorated and the name is under the buttons. There is an up button, a down button, and an off button. The fruit is cut before it goes into the Do-Lock. The fruit stars are tiny.

Two common errors encountered in metallurgical writings involve the figures: the identification/interpretation of microstructures and drawings and the words contained in a figure caption. I believe that such errors are common because the figures and figure captions frequently receive only marginal review. Proper review requires attention to details—all the details—not just the details in the text of the manuscript. Many reviewers spend too little time on the figures and simply work to improve the text. All too often we misinterpret the microstructures because we've failed to follow textbook descriptions of proper preparation techniques and failed to consult the references and standards describing the anticipated microstructural characteristics. Adequate assistance is available; it is simply not used.

Now, let us review Kate's Do-Lock. As far as I know, this is the first and only picture (macrograph, micrograph, or sketch) of a Do-Lock. Most of us would neither recognize a Do-Lock nor be able to identify any of its characteristic features. However, if asked, Kate would be able to further define the figure. She could establish whether the bright colors are clothes, skin, feathers, or metal; whether the Do-Lock is an animate or inanimate object; whether the magnification is 500× or 0.05×; and whether the figure is from the Do-Lock surface or is a cross section of its interior. She would know which part of the Do-Lock was sectioned and could determine the orientation of the cross section. Although virtually no reviewer would accept Kate's Do-Lock as part of a manuscript without obtaining more information, many reviewers will simply glance at a figure when an author describes or interprets a complicated microstructure. How often do we request that a figure or a figure caption identify phases, provide orientation information, establish the location of a metallographic cross section, and/or define the etching procedure? How often do we refer to published reference microstructures and fractographs to assure that an author's figure and/or interpretation are correct? Failure to review the figures can lead to errors in the manuscript, and when those errors occur, the fault should be attributed to the reviewer as well as to the author.

Fig. 2 Kate Lilly (Photo: Christopher Patrick)

Attention to detail is frequently the difference between success and failure. Reviews, whether of people, papers, or performance, are either bricks in the stairway to success or stepping stones in the path to failure. We spend much of our careers reviewing and being reviewed. Our successes and our failures will depend, to a large extent, on our willingness to examine our entire package—including figures, figure captions, tables, and even references—and to incorporate reviewer recommendations into our lives. How will you review the Do-Locks that you encounter, and how will you respond when your Do-Lock is reviewed?

Practical Failure Analysis, 2001, Vol 1(4), p 4–5

PROCRASTINATION: WHAT IS THE COST?

My wife Fran and I just finished a short course on successful financial strategies for retirement. The course covered insurance needs, investment strategies, wills, trusts, and tax minimization. On the first day of class, the instructor stated that the primary cause of financial failure during retirement was procrastination and emphasized the importance of saving for retirement when you are 30 or 40. I asked, "What do you do when you are 62?" and the instructor responded, "Work as long as you can!" He then proceeded to demonstrate, with elementary financial models, the cost of procrastination and convinced us that our financial strategies for retirement were initiated far too late. As the course progressed, we became painfully aware of the "things we should have done" and of the probable cost of our procrastination.

One of the important steps in determining the cause of a failure is organizing the failure analysis. Such organization is much like planning for retirement: recovery from a delayed start can be both difficult and costly. Unfortunately, there are no simple failure analysis models that demonstrate the cost of procrastination, and pressures often push for preliminary predictions even before the organization is completed. The words of a country western song, "I'm in a hurry to get things done. I rush and rush until life's no fun," describe an all too frequent approach to the failure analysis process. The rushing overwhelms the need for planning and organization and, surprisingly, pushes procrastination to the forefront. Forced business may be the primary cause of procrastination in failure analysis planning and organization.

During our thirties, Fran and I were busy with career, children, churches, houses, and friends, and retirement was simply something we would do tomorrow. Today it is easy to calculate the financial cost of our procrastination. (There were also benefits to procrastination—we've often told the children we would leave them memories rather than real estate—but the benefits of procrastination aren't the subject of this editorial.) Looking forward to a failure analysis, we become busy fulfilling customer requests, meeting management expectations, satisfying personal curiosity, and succumbing to the desire for the limelight. Organization of the failure analysis can quickly become something we will do tomorrow. In any high-visibility failure analysis, there are always pressures to start the analysis. Completing the tasks of planning and organization is seldom considered the actual start. Starting is making the first cut, taking the first fractograph, or measuring the first mechanical property. Even worse, starting may be making an educated guess as to the cause of failure. Procrastination over the task of organization can be costly in time, in dollars, and even in technical reputation.

A temptation that most analysts should try to avoid is the need to demonstrate expertise by guessing the result before the failure investigation is completed. An early, highly publicized, educated guess, especially a wrong or incomplete guess, may cause any of several unfortunate sets of circumstances:

- The guess may guide the failure analysis process, preclude proper organization, and compromise the results.

- The guess may be accepted as "fact," and the analyst may subsequently be forced to accomplish several unnecessary tasks simply to demonstrate that the "facts" were wrong.

- The customer may act on the guess and not wait until the analysis is completed.

Clearly, any of these circumstances may increase the time and resources required for the analysis and decrease the reputation of the analyst. Even a totally correct guess can damage a reputation by causing the customer to ask, "Why did the analysis cost so much (take so long) when you knew the answer the minute you looked at the failure?"

All too often we procrastinate on organizing the failure analysis and hide our willingness to procrastinate behind busywork and premature guesses. The busywork may be training, attending meetings, and assuring a proper quality assurance trail, all of which are important; however, we should recognize the cost of procrastination and using busy work to delay organizing the failure process. This cost is not something that is easy to measure, but it is relatively easy to recognize when we see redundant work, unnecessary analyses, and untimely completion. The code of ethics for most engineering societies includes some statement about acting in the public interest. When we procrastinate and subsequently fail to do things in a timely manner, are we really acting in the public interest?

Practical Failure Analysis, 2002, Vol 2(3), p 3–4

LOOKING BEYOND THE OBVIOUS

Jennifer Ellis Louthan is my niece by marriage to my nephew Bob, the mother of a baby girl, the owner of a new house, and an extraordinary artist. Several weeks ago, my wife Fran and I visited Jennifer and her family to help celebrate the arrival of the new baby. We were surprised and delighted to also receive a print of some of her artwork. The print was of a patriotic scene she painted shortly after the September 11 attack on the World Trade Center. A framed print of the painting hung on their den/entrance hall wall and captured me as soon as I saw it. The painting had an American flag in the background and an eagle in flight in the foreground, and the eagle's beak and talons were quite apparent. There were also arrows and an olive branch. I wanted a print for my office wall but tried not to let my desire be obvious. As we were leaving, Jennifer asked whether we would like to have a print and described some of the meaning behind the painting. I was amazed at the details I had missed. Before she discussed her painting, my description would have been similar, if not identical, to the one just made. I had only looked at the obvious.

The task of many failure analysts is to look beyond the obvious: to "spy out" what is half there, to see features and patterns that are easy to overlook, and to recognize the important, the critical, and the difficult to reveal. The article "General Practices in Failure Analysis" (*Failure Analysis and Prevention,* Vol 11, *Metals Handbook,* 9th ed., ASM International, 1986, p 15–46) emphasizes the need to look beyond the obvious without ever making a direct statement to that effect. For example, the article states that an "analyst should decide if photographs of a failed component or structure are required" and then suggests that "a failure that appears almost inconsequential in a preliminary examination may later be found to have serious consequences." This suggests that photographs are generally necessary. I don't believe I've ever heard a failure analyst say, "We certainly took too many photographs," but I have often heard, "I wish we had a photograph of... " It is easy to overlook important features, and without a photographic record, evidence we should have had may be missing. My preliminary examination of Jennifer's painting had missed a significant fact: the eagle had dropped the olive branch. If the failure analyst doesn't look beyond the obvious, significant facts may be missed, and a conclusion and/or recommendation that should have been obtained readily may remain hidden.

The general practices article also states that "the examination should be performed as searchingly and effectively as possible, because clues to the cause of breakdown are often present but missed if the observer is not vigilant enough." Vigilance in a failure analysis takes many forms but haste is generally not one of them. Last fall I heard a rather senior (politically correct word for *old*) failure analyst state that the scanning electron microscope (SEM) was the best and the worst tool available to the modern failure analyst. It is certainly an excellent tool for fractography and for chemical analysis with appropriate attachments: however, a rush to examine the details of a fracture surface can destroy lots of evidence. The general practices article recommends that "fracture surfaces should be cleaned only when absolutely necessary" and further recommends cleaning with cellulose acetate replicas to preserve the debris for future examination. Unfortunately, there is a tendency to rush to the SEM, and the rush frequently involves chemical (water, alcohol, etc.) cleaning and sectioning the failed component. We have forgotten that "the amount of information that can be obtained from examination of a fracture surface at low-power magnification is surprisingly extensive." If I had examined Jennifer's painting carefully, I might have noticed that the olive branch had nine olives and that the eagle was carrying eleven arrows thus illustrating the date 9-11. Too often we fail to count and measure fractographic features, crack lengths and openings, and other distinguishing characteristics of the failure. We may even fail to look at both sides of a fracture in spite of our knowledge that the examination of matching surfaces may be necessary to distinguish the stress state that caused the fracture to develop.

There is a checklist of questions that the "General Practices in Failure Analysis" article presents to act "as an aid in analyzing the evidence derived from examinations and tests [conducted in the failure analysis] and in formulating conclusions." The answers to some of the questions are apparent, while other answers are difficult to obtain. Some of the questions may not be relevant to a specific failure; however, a review of the list may help us look beyond the obvious and find important contributing factors to a failure process. These factors may have remained hidden and not discussed in the recommendations if we had been satisfied with a simple answer about what happened (fatigue, hydrogen embrittlement, stress corrosion, or overload). The solution for many problems must involve answering the question *why* rather than simply describing the mode of failure. In Jennifer's painting the eagle was carrying a single arrow in his beak, signifying unity. As we conduct our next failure analysis, let's try to look beyond the obvious and provide the evidence necessary to develop a single unifying solution to the problem.

Journal of Failure Analysis and Prevention, 2004
Vol 4(2), p 3–4

AWARENESS AND GRANDFATHER CLOCKS

We were sitting around the dinner table at a Cracker Barrel restaurant, celebrating Rea's birthday. Rea is my wife Fran's sister, and most of the family had gathered in Columbia, S.C., for the celebration. No one in the family lives in Columbia so everyone had to travel, but in today's world travel isn't just okay, it is expected. Fran's birthday present to Rea was a clock. It was a simple battery-powered clock that probably cost less than 15 dollars; however, Rea needed a clock and the clock became a topic of conversation. In the midst of the conversation, Fran's brother Tom asked if we knew the Roman numerals used on the faces of grandfather clocks. Everyone said that they did, but everyone was wrong. Somehow, after 70+ years of seeing grandfather clocks, I had not seen that the numeral four on the clock dial is now not IV, as the correct usage of Roman numerals would suggest, but is IIII. At first I didn't believe Tom. Why would anyone, much less an entire industry, use the wrong symbol for the numeral four? I still haven't found a satisfactory answer to that question, although I examined enough clocks to assure myself that IIII is generally used where the symbol IV should exist. Fran's watch, a clock at the Royal Observatory Greenwich, and a platform clock at King's Cross railway station in London all have the IIII symbol for four.

A Google search for the history of clock faces with Roman numerals provided numerous suggestions as to why the IIII is used. Symmetry on the clock face, ease of casting the numerals (because when IIII is used there are ten Is, two Vs, and two Xs on each side of a central core), to avoid using IV (which was an expression for the god Jupiter), and a decree of a king were all suggested, but there was no real consensus. However, the problem I had with Tom's revelation is not with the history, the tradition, or even the ease of casting the numerals. My problem is that after 70 years of looking at grandfather clocks, I'd never been aware that the symbol IIII was used. This lack of awareness became an even bigger concern when I found that questions about the clock face are frequently ask. I found, during the Google search, a section listed as "FAQ: the Roman Numerals on Clock Faces." I assume the letters stand for Frequently Asked Questions.

How can a grown man who has spent much of his professional career examining microstructures, fracture surface topographies, and worn, corroded, or distorted components have missed something as apparent as the use of IIII instead of IV to represent the numeral four on grandfather clocks? The only answer I can find to this question is a lack of awareness, and this answer raises the question, "What else have I missed?"

Awareness, attention to detail, and recognition of unexpected features are key elements in any high-quality failure analysis. Experience is generally an excellent instructor. Industrial salaries typically increase with experience, and no one wants to be a brain surgeon's first patient. Yet there is an old saying that "familiarity breeds contempt," and in some cases familiarity has led to a lack of awareness. When you looked at the Roman numerals on a clock face, did you miss the IIII?

The fractographic characteristics typical of fatigue, hydrogen embrittlement, chloride stress-corrosion cracking, creep, and other common age-related materials degradation processes are well recognized and relatively easy to distinguish from tensile overload, torsion, or tearing. The metallographic indicators of sensitization can be found by exposing stainless steels to oxalic acid, and an energy-dispersive analysis of X-ray attachment to a scanning electron microscope can be used to map the distribution of elements on the surface of an iron, copper, or aluminum alloy. Many failure analysts routinely use materials analyses technologies to enhance their understanding of a failure. Metallographic and fractographic examinations are standard components of many, if not most, metallurgical failure analyses. The question that should be raised by the grandfather clock face is "Do the examinations ever become so routine that an unusual feature is overlooked?" This only happens when the examiner assumes, after a quick glance, that all the characteristics seen are as expected. When I examined the face of a grandfather clock, I assumed the configuration of all the numerals. I actually was correct for over 90% of the numerals, but >90% accuracy allowed me to totally miss the IIII.

Small, unnoticed features that resemble what is expected are easy to miss. Let us resolve, as failure analysts, that we won't miss any indication simply because a quick glance seems to present the image that we expect to find. Don't simply say you certainly haven't missed an obvious alteration; after all, you may have looked at grandfather clocks for years without being aware that the IV was altered.

Journal of Failure Analysis and Prevention, 2011
Vol 11(2), p 77–78

TEAMWORK AND RESPONSIBILITY

ILLUSIONS

Many years ago, my family and I, along with several other family groups, were hiking on the Appalachian Trail. Our destination was Angels' Rest, a mountaintop just outside Parisburg, Va. The afternoon hike was a round-trip excursion from Parisburg, and, although the trail was occasionally steep, the trip was in no way a mountain climb. When we reached the top, Rich McNitt turned to my daughter Amy, who was about five years old at the time, and said, "That sure was fun, wasn't it?" Amy's response surprised me. "It may have been fun for big people, but for little people it was scary!" That was one of the first times I realized a rather obvious fact: a shared experience does not necessarily stimulate the same responses in two different individuals. Long before the hike, life should have made me aware that individuals can see the same things, hear the same things, and even experience the same things and reach very different conclusions.

My high school basketball coach, Lawrence (Burrhead) Bradley, was one of my mentors and heroes. He was an old-time coach who thought that you couldn't run too much, practice too hard, or study too long. Blood, sweat, and tears were ordinary and appeared in virtually every practice, especially during tryouts for the high school team. In the fall of 1955, about 20 boys tried out for varsity basketball at Graham High School. Only 12 made the team. One of the boys who did not make the team described Mr. Bradley as a mean, unfair man. In the fall of 2000, at the 45-year reunion for the class of 1956, the boy of 1955 had become a man, but he still held the opinion that Mr. Bradley simply required too much of an individual. None of the team members shared that opinion. To a man, we believed that Coach Bradley taught us that effort, discipline, goal setting, and practice were keys to success. I believe that if I'd demanded as much from myself on the job as Coach Bradley demanded from me on the basketball court, my career achievements would have been better than they are. I experienced the Graham High School basketball tryouts and view them as a positive influence to my life, while another boy viewed the same set of variables as cruel and unusual punishment. One, or perhaps both, of those views is an illusion.

During the O.J. Simpson trial, we were all hearing the same radio broadcasts, reading the same newspapers, and seeing the same trial on television. One set of Americans concluded that Simpson was as guilty as a fatigue crack, while another set of Americans concluded that he was as innocent as a newly formed single crystal. Clearly, one of these views is an illusion. Illusions are all around us. They follow us to work, they attend our meetings, and they interfere with our well-formed conclusions.

We recognize that when two athletic teams are competing, the events that are good for one team are generally bad for another, but we rarely extend this recognition into other areas of our lives. The same event may have very different effects on individuals, depending on their choice of teams. Life experiences and circumstances place each of us on a team—some on a management team, some on a research team, and some on a production team. Some teams analyze the results of another team's activities, some teams try to improve another team's performances, and some teams simply try to profit from another team's misfortunes. We are all on one or more of these or other very similar teams. Unfortunately, the goals, the experiences, and the expectations of each team are different and the team's view of its position or importance may be an illusion.

Several months ago, a corporation was working on a project that involved projecting, then maintaining, the useful service life of a rather complicated system. The production team had experienced problems with the system and wanted to know what it would take to ensure that the system would not fail in the near future. Their focus was on having the ability to continue production, and, in their eyes, long-range planning was looking ahead for two weeks. The management team wanted assurances that, regardless of what it took to save the system, the cost of ensuring that the operation could continue would be within budget and production schedules could be met. Their focus may have been more long range than that of the production team but only because there were several months left in the fiscal year. The research team wanted to obtain the information necessary to define potential system failure scenarios and establish the operational windows that precluded failure. Their focus was clearly long range and included tests and analyses that would not be completed until after the scheduled shutdown of the production system. The oversite team didn't give a "stress-corrosion crack" about production, schedules, cost, or failure; they wanted to be assured that environmental quality was maintained and that whatever was done met all applicable regulations and standards. Each team heard the same presentations, saw the same viewgraphs, studied the same data, and had discussions with the same individuals. However, the conclusions of each team were different. When team conclusions were presented, the production team thought that the management team was too tight with a dollar, that the research team had inhaled the offgases from too many molten lead pots, and that the oversite team had majored in confusion and obtained Ph.D.s in disruptive processes. No team could agree with the conclusions of any other team—and all the teams were amazed that the other teams did not agree with their conclusions. Somehow, we frequently think that there should be no illusions at work, when we already know that illusions are basic to the human experience.

The next time you participate in a meeting and another participant uses your team's data to reach a conclusion that you cannot support, try to think, "ILLUSION." Most of the team members I've met aren't dumb, even when they disagree with me. Their sets of life experiences simply differ from mine. Frequently, I've found that illusions can be altered and reality can emerge when discussion is uninhibited and the various players recognize the importance of attitude to the meeting's success. Attitude was something that Coach Bradley always emphasized. He said that performance was never more than 50% dependent on ability and never less than 50% dependent on attitude. Your attitude can eliminate or reinforce someone's illusion; however, if you want your team to win, you must also recognize that the illusion could also be yours.

Practical Failure Analysis, 2002, Vol 2(6), p 3–4

AVOIDING THE BLAME

One summer our family—children and grandchildren—gathered at Claytor Lake to celebrate the Labor Day weekend. The grandchildren's ages ranged from less than one to thirteen. Their heritage has led to "type A" personalities, and, as a child reaches the magical age of two, each assumes that he or she has to become the family leader. Some of the joy of a family gathering was watching Ty (2) and Hunter (4) try to force Emily (13) and Austin (8) to bend to their will. However, in spite of the constant positioning for leadership, they get along very well and the gatherings are mostly harmonious—until something is spilled, broken, or misplaced.

We were roasting marshmallows, toasting graham crackers, and heating chocolate bars to make "s'mores." Marshmallow roasting takes place in the evening, when everyone is tired and we are transitioning toward bedtime. We build a small fire in the chiminea that sits at the edge of the deck. When the coals are just right, the grandchildren compete for a position around the fire, and the parents worry about the hot coals and frown at me for encouraging such a dangerous activity. The grandchildren love it! Sometimes I think that they enjoy the marshmallow roasting more than the s'more eating.

S'more building and eating is a very messy activity: chocolate and marshmallow get everywhere—especially on the children, but they wash, so we don't worry. However, my wife Fran and I do try to keep the ingredients off the wooden deck and suggest frequently that the s'mores be built and eaten on the grass. During the middle of the Labor Day marshmallow roast, I spied two freshly roasted marshmallows smashed into the deck. The location, coupled with the thumbprints in the middle of each smashed marshmallow, suggested that the smashing was not accidental, so I asked, "Who did this?" The immediate response from each of the five grandchildren who were not being held by a parent was "Not me!" Kate (5) added, "I don't even know who did it." I didn't try to determine the culprit and didn't think of establishing the reason(s) for the smashing. (Grandfathers can't perform an unbiased root cause failure analysis that involves their own grandchildren.) However, I was reminded of a failure that I analyzed several years ago when a colleague and I basically asked a manufacturing division, "Who did this?" and the response from each element of the division was "Not me!"

We were called to investigate the occurrence of compressor head bolt failures in new systems that were "identical" to old systems that did not experience such failures. Our analysis was only hours old when we became relatively certain that the bolt fractures were due to delayed failure hydrogen embrittlement. We set up an evaluation to establish what had changed in the bolt fabrication and plating processes. We needed answers to questions involving material specifications, thread rolling, heat treatment, cleaning, electroplating, bake-out, and even bolt design. To obtain the needed answers, the managers of five procurement/manufacturing groups involved in the production of the bolts were called to a meeting. Each manager made a short presentation describing his or her group's roles and responsibilities. The atmosphere surrounding the meeting was, at best, not relaxed.

The fact that bolt failures began to occur after many years of success suggested either single or collective alterations in the bolt production operations caused, or at least contributed to, the failure process. Therefore, my colleague and I asked each presenter, "What recent changes have been made in your group's production operations?" The response of each manager was exactly the same: nothing had changed. After all the presentations were completed and we had experienced several hours of post-presentation discussion, my colleague stood and shocked everyone, including me, by stating: "Mac and I have found the problem. The phase of the moon, coupled with an adverse air temperature and relative humidity, caused the problem and there is no reason for any further discussion. The phase of the moon will be correct in two days, so you can just pay us and we will head home."

The room was relatively tense for a moment or two before the manager who had hired us stated, "You can't be serious; the phase of the moon has no effect on delayed failure hydrogen embrittlement." My colleague responded that he wanted to agree with the manager, but because he had been absolutely certain that the onset of delayed failure hydrogen embrittlement was caused by a change, or several changes, in the production process, and the meeting had demonstrated that nothing had changed, the phase of the moon, which does change, must have been responsible. This statement relaxed the group, and slowly we began to hear that every group had made "relatively minor and totally irrelevant" changes to improve their portion of the production process. After several more hours of open and frank discussion, the source of hydrogen uptake and reasons for postplating hydrogen retention began to emerge, and by the next day the problem was solved. A plan to replace potentially susceptible head bolts was then developed and production was resumed. In this case, a root cause failure analysis was completed successfully because the production teams decided that solving the problem was more important than avoiding the blame. Unfortunately, when a failure occurs or a problem arises, many of us have a tendency to say, with conviction, "NOT ME!"

Fran used to say that "Not Me" was responsible for so many problems in our household that she was going to punish him severely if she ever found him. How frequently have our attempts to avoid blame delayed, inhibited, or otherwise adversely affected our participation in a failure investigation? Have we trashed potential production improvements because we didn't want to be shown to be the current bottleneck? I'm absolutely convinced that marshmallows on the deck were smashed by one of the grandchildren, even though I didn't question my adult children or their spouses. Smashing marshmallows is childish, but then, so is the tendency to try to avoid blame. Let's all try to accept the responsibilities for our actions; after all, we are adults.

Practical Failure Analysis, 2003, Vol 3(1), p 3–4

ASKING FOR HELP

One of the problems with becoming an experienced metallurgist (another expression for *old* metallurgist) is that some of your talents deteriorate. I spent the better part of an afternoon attempting to rationalize an accumulation of data on hydrogen diffusion into a cylinder. The rationalization required my using differential equations and Bessel functions. There was a time when this type of arithmetic would have been no problem, but this old metallurgist had faced neither differential equations nor Bessel functions in decades. The afternoon was spent attempting to relearn long-forgotten math. My recall was far from immediate, and I was dog-tired when the day was through. Fortunately, the weekend provided time for rest and recovery.

During the weekend, my wife Fran phoned our daughter Amy. Fran's call interrupted a mother-daughter project that included painting ceramic dishes. During the call, Amy suddenly said, "I have to go, Mom— Kate just spilled paint all over the kitchen!" Fran waited an hour or so, then called Amy to see how things were going. Amy was calm, the kitchen was clean, and Kate wanted to talk to her Mamaw. During this grandmother-granddaughter conversation, Fran asked Kate, "What happened?" Kate explained that she spilled the paint when she attempted to remove the top from a new container of blue paint. Fran's question, "Why didn't you ask Mommy for help?" was answered with the statement, "It's hard to ask for help when you are five years old." When Fran shared Kate's answer, I thought about differential equations and Bessel functions. There are numerous skilled engineers in the materials group at Savannah River Technology Center, and I had failed to ask for help. Why? It is hard to ask for help when you are 63 years old!

Asking for help, especially asking for help when faced with a problem that you think you should be able to solve alone, is difficult. It is difficult at any age. Why? Since that weekend I've questioned my reluctance to ask for help. Fran claimed that this reluctance is a human characteristic that girls outgrow as they approach maturity and males retain throughout their lives. The "fact" that men are typically unwilling to ask for directions was stated as a proof for her claim.

There are at least three reasons why asking for help is difficult. First, much of our training and education focused on self-reliance, and asking for help becomes tantamount to admitting personal failure. Second, asking for help places us among the needy, and we don't want to be perceived as "a recipient of charity." Finally, to ask for help, one must admit, "I can't, but perhaps they can." Each of these reasons plays on our self-confidence and self-image and causes many of us to face failures alone when good help is only a phone call away.

Successful failure analysis, especially when the failure involves a complicated system, generally requires the application of several engineering disciplines and may also involve several areas of expertise within any given discipline. This observation suggests that a multidisciplinary approach will generally be the best approach to a failure analysis. Unfortunately, most of us are not experts, or even competent, in the wide variety of disciplines that may be applicable to the failures we face. Therefore, true success requires our asking for help. However, asking for help is only the first step. We must be willing to accept, or at least listen to, the advice that our request for help may bring.

Initially, I avoided asking for help with the Bessel function problem by spending an inordinate amount of time attempting to relearn old skills. Even then I wasn't confident that my rationalization was correct so I asked someone to review my work. There is a trick frequently used by engineers and scientists who, even though they are uncertain of their analyses, don't want to ask for help. They simply ask for a review of the work but don't point to the areas where the work is weak. Perhaps unconsciously, the users of this trick hope that the reviewer will bless the work and miss or overlook any weaknesses. When this happens, the perception of personal failure is avoided. Unfortunately, so is the application of a necessary expertise. This trick may avoid asking for help, but it may also minimize the success of the failure analysis. Additionally, subsequent failure prevention methodologies may be limited or even incorrect.

There is another technique for asking for help that avoids admitting that we are among the needy. We need help but only ask for confirmation: "Please look at this and see if everything is all right." We ask for advice, but too frequently, when advice is offered, we argue with the advisor. This technique creates the perception that, even though we weren't among the needy, we were willing to listen when "unneeded" help was forced upon us. The use of this technique also avoids the perception of personal failure because we never had to admit that we lacked the expertise to address all aspects of the problem.

Finally, there is another technique to avoid admitting that we can't address some aspect of the analysis. We say something equivalent to "I'm very busy right now. Could you address this aspect of the problem?" The busy schedule scenario creates the perception that we could do the analysis but are tied up with more demanding problems. This technique of avoiding asking for help does focus the needed expertise directly on the problem; unfortunately, the downside is that we have implied that "your time isn't as valuable as mine, so you should solve this minor aspect of the problem."

The successful failure analyst must learn to ask for help. Such requests are not signs of weakness but demonstrate that the requestor recognizes the need for additional expertise and understands that a multidisciplinary approach may provide a better solution. It is "hard to ask for help when you are five years old," but let us all mature to the point that we have the self-confidence and wisdom required to ask for help when help is needed. After all, each of us has an extensive network of co-workers, friends, and acquaintances who would help, if asked. The failure to ask for help may cause you to spill paint on your kitchen floor.

Practical Failure Analysis, 2003, Vol 3(3), p 3–4

BLOCKING THE LANE

Some time ago, I participated in a brainstorming session designed to solve, or at least provide a pathway for addressing, a recurring problem. About twenty people assembled in a conference room, and after one person described the problem, the participants offered potential solutions. "Out-of-the-box" ideas were encouraged. Participation rules stated that there were no dumb ideas, allowed "piggybacking" on another participant's idea, and delayed evaluation of the ideas until brainstorming was completed. About halfway through the meeting, I thought of something one of my children told me approximately eight years ago.

The men in my immediate family had entered the "Hoop-It-Up" three-on-three basketball tournament in Richmond, Va. I was 57, my two boys were 32 and 29, and my son-in-law was 27. Substitutions were allowed, and even though it was a three-on-three tournament, virtually every team had a four-man roster. Although we fancy ourselves a basketball-playing family, the competition was rough and winning was difficult. We were facing elimination, and because winning is important, we were working hard to salvage the game. Twice I had moved into the lane, received a pass, and, rather than shooting, passed the ball back to one of the boys. My son Keith called time-out. As we strategized in the huddle, Keith said, "Dad, when you get the ball under the bucket, shoot." I responded that it was hard for me to make a move to shoot because my old muscles didn't like to turn against my momentum. Keith immediately said, "If you can't shoot from the middle, stay out of the lane. You are blocking the rest of us." Even though I didn't do it again, we lost the game and were eliminated from the tournament. Although I had not thought about it until the brainstorming session, there is a question that I should have asked: Was that loss a result of my blocking the lane?

Years ago, I took a short course that had a "lane blocker" as the professor. He intimidated the class by insulting the first two students who asked questions. Virtually all the students were working professionals, and after the first two questions were addressed with insults, no one was willing to ask the third question. The short course lasted three days without another question from the class. I did learn something from the course but not nearly as much as I should have learned. The lane blocker had prevented questions and, by preventing questions, had minimized discussion. Ridicule and intimidation are effective lane-blocking techniques.

We have all met someone whose physical or emotional actions block the lane. Not only do they keep others out of the lane, the lane blocker frequently fails to shoot, even when he or she has the ball right under the bucket. Lane blockers include:

- The Complainer
- Those who are "too busy" to help
- Failure analysts who reach conclusions without all the necessary facts
- People who dominate a meeting
- Anyone thinking that he or she has all the answers

Although most would never admit it, lane blockers operate under the perception that "my time (idea, job, talent, etc.) is more valuable than yours." Unfortunately, most of us become lane blockers on occasion.

The lane blocker in one of our brainstorming sessions was a young fellow who thought he had all the answers. He talked and talked, repeating his ideas by rephrasing his presentation. He dominated the meeting but failed to provide any significant contributions. In the end, the brainstorming session failed to provide a solution to the problem, partially because of the actions of the young fellow. Generally, anyone dominating a group discussion becomes a lane blocker, and, given the opportunity, we all tend to steer clear of lane blockers. We like to be surrounded by lane *clearers*—people who clear a pathway for ideas and provide others the opportunity to shoot. Lane clearers are a necessary part of a winning team.

My boss is a lane clearer. He firmly believes that a significant part of his job is to provide everyone in the materials section an open lane for work. He sets high standards, provides good ideas, and champions excellence. He listens and responds to concerns, fosters cooperation, and insulates us from the dragons of bureaucracy. He shares ideas, supports his people, and creates a stimulating work environment. People enjoy working in his organization primarily because he helps make good things happen.

When facing a problem, attending a meeting, or simply discussing our work, we generally choose one of several positions. We become a lane blocker, a lane clearer, or a shooter. Talking without thinking, finding fault, and itemizing potential problems can be done with minimal effort. These lane-blocking techniques make meetings ineffective, hurt morale, and stifle innovation. Lane clearing is as beneficial as lane blocking is harmful. A lane clearer may elucidate a success pathway, correct an inappropriate situation, or simply stimulate someone's thought process. Lane clearers seldom seek credit for their work. They are team players and are significant assets to their organizations. Every team needs at least one lane clearer. Shooters are the keys to success, especially if the shooter works in cooperation with a lane clearer. Shooters are the innovators, the problem solvers, the idea people, and, because they are shooters, are generally very busy. Shooters are decisive, willing to take a risk, and look for reasons for success rather than reasons for failure.

My boss, in addition to being a lane clearer, is a shooter. He runs with ideas, takes technical risks when the potential reward is high, and focuses on the pathways to success. Our organization is growing, expanding its capabilities, and extending its reach. Unfortunately, even while this is happening, the lane blockers are discussing the potential for failure and the challenges that our organization faces.

As I look at my career, I find that I've been a lane blocker at times. I've talked when I should have listened, complained when I should have had a positive focus, and spent too much time discussing how failure might occur when success was just around the corner. How about you? Are you a lane blocker, a lane clearer, or a shooter? Remember, when you are in a situation where you aren't going to shoot, get out of the lane!

Practical Failure Analysis, 2003, Vol 3(6), p 3–4

FAILURE PREVENTION AND DESIGN

IN DEFENSE OF THE METALLURGIST

The metallurgical/materials engineer is frequently blamed when a material fails in service. "Two thousand years of metallurgy and the profession remains a black art!" "Why can't the metallurgist select appropriate materials?" and "Hasn't the materials industry developed anything that works?" are statements made and questions asked after a failure. Before the failure occurs, the statements and questions are vastly different: "We don't need a metallurgical review of this project." "Isn't there anything cheaper we could use?" and "Could you give us an evaluation of the materials we selected within the next two hours?" It is very difficult to assure that metallurgical and materials evaluations are a regular part of the design process for most systems. My experience suggests that materials considerations are often a backfit that is too little and too late to assure project success.

Materials issues do not receive appropriate attention in many major projects primarily because a strong interface between the design and materials functions is not formally established. Deficiencies in this interface may be reflected in many ways, including premature failures in systems and/or components because of improper materials selection and/or specification. An equally poor alternative to system failure is system over-design to assure against a materials degradation process that does not and should not occur. The lack of an effective interface is so common that when failure analysts evaluate the reasons for failure, six fundamental causes are established: deficiencies in design, improper materials selection, defects in materials, improper materials processing, errors in assembly, and inadequate service.

Clearly, three of the six causes are materials related. A closer inspection shows that materials also play a role in two others. Errors in assembly, for example, include improper welding, and inadequate service includes operation outside the anticipated process or flow sheet windows. Welding is clearly a materials technology, and operation outside flow sheet windows may force a material to perform under inappropriate and/or unanticipated conditions.

The importance of the materials-design interface is further illustrated through a breakdown of the typical causes of corrosion-induced failures. Only 8% of corrosion-induced failures result from a lack of awareness of the corrosion risk, and 10% result from a material not performing as it should. These percentages are lower than the 17% attributed to unforeseen operating conditions, 17% to improper protection of materials (coatings, lubricants, etc.), and 11% to poor process control. Additionally, 19% of corrosion-induced failures result from design faults—faults that presumably would have been corrected by proper attention to the materials-design interface.

The fact that a large fraction of industrial failures involve materials degradation is often ignored for two reasons. First, the team responsible for project design and construction is not the team responsible for operation and maintenance of the facility. Second, national consensus codes and standards rarely include specific provisions for age-related degradation processes other than uniform corrosion and fatigue.

However, if a project is to meet operational goals, serious attention must be given to materials qualification, selection, and validation. In fact, validation processes may be vital to the ultimate success of many projects, because the qualification efforts frequently are in parallel with, rather than prior to, project design. Based on these background observations, a materials-related program will play a key role in the success (or failure) of the long-term operational readiness of most major projects. Unfortunately, the introduction of metallurgical considerations to a project is frequently an afterthought, and the anticipated role for materials is secondary. Additionally, a reduction in materials costs is generally considered a cost savings rather than an increase in the potential for failure.

Most metallurgists are fully aware of the need for increased materials considerations; however, many designers are not aware of the importance that materials play in long-term project success. Perhaps, just perhaps, if the designers could become more aware of the real causes of corrosion-induced (and other) failures, the materials and design functions would become fully integrated early in the design process resulting in fewer failures.

Practical Failure Analysis, 2001, Vol 1(3), p 3

HIGHLIGHTING FAILURE PREVENTION

The key findings of a Federal Highway Administration study on the cost of corrosion to the economy of the United States are summarized in the tech brief FHWARD-01-157.* This report concludes that the direct annual cost of corrosion is 3.1% of the U.S. gross domestic product ($276 billion in 1998 when the GDP was $8.79 trillion) and that the indirect costs are at least equal to the direct cost. The Tech Brief is well worth reading and several of the observations are, at first glance, surprising. For example, the direct cost of corrosion of drinking water and sewer systems was, at $36 billion, higher than the cost of corrosion in the entire transportation sector, $29.7 billion. However, there are 876,000 miles of municipal water piping and the U.S. sewer systems release over 41 billion gallons of wastewater per day. Further reading also shows that part of the estimated $36 billion annual cost of corrosion includes "the cost of replacing aging infrastructure and the cost of unaccounted-for water leaks." *Sure, blame it on corrosion and don't even suggest that fatigue or other failure mechanisms might be involved.* This is almost like blaming a leak in a J-tube on improper application of the chrome plate. I've never been able to understand why the chrome plate is on the outside of the J-tube when corrosion generally initiates on the inside. My wife says that the plate is on the outside so the consumer can see it, know that it is there, and be certain that he or she is buying a chrome-plated sink trap. Future performance isn't nearly as important to the sale as presentation and packaging. Perhaps she is correct—

*Copies are available from the National Technical Information Service, 5285 Port Royal Road, Springfield, VA 22161 (the full report is FHWARD-01-156).

maybe it's all a matter of making the sale, but then, maybe, just maybe, there is real reason for the chrome plate. In any event, I did not intend to criticize the Tech Brief; I simply wanted to use some data from the brief to discuss the value of failure prevention.

The study estimated that there are over 200 million vehicles registered in the United States. These vehicles, at an average assumed value of $5000, represent an investment of approximately $1 trillion. The total annual direct cost of corrosion for these vehicles is $23.4 billion, or approximately $117 per vehicle. The direst costs include increased manufacturing costs related to corrosion avoidance, repairs, and maintenance necessitated by corrosion-induced degradation and corrosion-related depreciation. The annual cost of repairs, maintenance, and depreciation ($20.9 billion) is approximately $104 per vehicle; thus, the annual cost of corrosion avoidance is approximately $13 per vehicle. I believe that the return on this "cost of corrosion" is remarkable. The first three cars that I owned were manufactured in 1958, 1963, and 1968. All three failed because the body rusted away. I even acquired a 1965 compact car from my brother when he thought that the rusted body offended both fish and fishermen. My brother had gotten the car when my mother decided that it was too rusty for her to drive. The point is, the motor was working fine, but the body had basically rusted away. My last three cars were manufactured in 1985, 1994, and 1995. None of these cars show any body rust. The 1985 car was traded for the 1994 automobile in 1996, and at that time, there was no sign of rust on the body. The 1994 car and the 1995 truck also show no signs of rust. My experience suggests that modern cars, if you can consider my old relics modern, simply don't rust. Service lifetimes are now predicated on something other than a rusty car body or the floorboard falling out. In my case, and I hope in the case of most other vehicle owners, the $13 per vehicle portion of the annual "cost of corrosion" translates directly into savings. Although some of the savings may be monitored, others are very difficult to evaluate, much less quantify. Placing a real, quantitative value on cost avoidance by the prevention of corrosion-induced failure is one of those difficult things that "the bean counters" tend to ignore. Thus, the savings from corrosion avoidance are easy to ignore, while the cost of corrosion avoidance is easy to quantify.

One element of the Federal Highway Administration's cost of corrosion study is the "cost of additional or more expensive material used to prevent corrosion damage." The analysis apparently compares the costs associated with the use of corrosion-resistant alloys and/or protective coatings with the cost of noncoated, plain carbon steel and attributes the difference to the cost of corrosion. This is exactly the thought process ascribed to "the project" when a materials engineer or failure analyst suggests the use of a better material or an improved design. Too often, immediate manufacturing/production costs override life-cycle costs, and components, systems, and structures are not improved, even when improvements are relatively easy. The lack of improvement results in an immediate cost saving (or cost avoidance), and "the bean counters" simply do not have the cost data necessary to support the use of a more costly alternative. Corrosion-resistant materials sit on the shelves, coatings remain unused, and new designs stay on the drawing boards.

I'm not sharp enough to establish a process that determines and highlights the cost savings that result through the application of well-established failure prevention technologies. However, I am glad that the vehicle manufacturers decided to spend that $13 per vehicle per year to avoid corrosion. This is one cost that I'm delighted to have paid and simply hope that similar investments are soon made in other arenas. Certainly, the automobile industry has a lot to say about highlighting failure prevention. We should take the time to listen and try to learn from their accomplishments.

Practical Failure Analysis, 2003, Vol 3(2), p 3–4

FOLLOWING THROUGH

A newsletter from Randy Frye, the senior pastor of our church, recently emphasized that the practice of following through is vital to a successful life. Randy is a sports enthusiast and was, according to at least one sermon, "a relatively good college athlete," so he used athletics to introduce the points in his letter. The golf swing doesn't stop when the club contacts the ball; the arm/hand movement on a basketball shot doesn't end when the ball leaves the hand; and a football tackle has just begun when the tackler contacts the ball carrier. The success of these endeavors depends on following through and wrapping up. My wife Fran was reading the letter when I received a call from Jo Hannah Leyda, managing editor of the *Journal of Failure Analysis and Prevention.* We discussed several things that I needed to accomplish before she could continue her work on the issue in question. (Said gently, I was behind schedule.) Fran was overhearing my end of the conversation, and after Jo Hannah and I had finished talking, Fran handed me Randy's letter and suggested that a habit of following through might help me in meeting my commitments to the journal. As I read the letter, thinking about the journal, I renewed the realization that following through and wrapping up are essential elements in the practice of failure analysis and prevention.

Failure to follow through in the rendering of service is one of the fundamental causes of failure. Oiling, greasing, cleaning, painting, inspecting, aligning, and many of the other aspects of service are easy to put off—easy to avoid, and the avoidance of service may even be considered a cost-saving item by shortsighted management. Most of us have investigated failures that "occurred" months after the first leak was detected. Many of us have completed an analysis and given a verbal report on the cause of failure weeks, months, and even years before the formal report on the

Failure to follow through in the rendering of service is one of the fundamental causes of failure.

failure was prepared. The enthusiasm that existed at the beginning of the analysis wanes. After being informed of the cause of the failure and receiving recommendations for preventing future failures, the customer's interest in, and support for, following through and wrapping up frequently declines. I've been involved in numerous failure analyses in which the customer was basically uninterested in my wrapping up the investigation: the customers had all the information they wanted. On more than one occasion, the customer was totally focused on short-term success, and long-term recommendations were essentially unwanted. Following through under these circumstances may be very difficult. However, following through is always important. Perhaps, just perhaps, following through and really completing a failure analysis is the single most important event in the analysis.

Following through involves preparing a written report that includes recommendations and, whenever possible, sharing emergent information with the public. All engineers have an obligation to protect the public, and when over one percent of the gross national product of most developed nations is lost because of improper service, the failure analysts must not be properly sharing the information they have gained. The failure analysis community knows, better than any other group, the consequences of improper service, the consequences of failing to pay attention to "lessons learned," and the consequences of operating industrial systems with a focus on "success in my time." However, in spite of our experience, in spite of our knowledge, and in spite of the insight we have gained, we are often guilty of failing to follow through.

Several months ago, I was trying to obtain a commitment for a manuscript only to have the prospective author state that he would prepare his paper after he received a written copy of an oral report I gave at a recent technical meeting. Oral reports are generally enjoyable experiences but do not contribute significantly to the knowledge base required for failure prevention. Referencing an oral report is difficult, and handing an oral report to someone needing the information is virtually impossible. Publishing and making recommendations for failure prevention are the "following through" portion of the failure analysis process, going far beyond simply presenting an oral report.

Failure prevention requires knowledge, insightful recommendations, and a willingness to incorporate lessons already learned. Knowledge is only useful to the public when shared. Therefore, to meet our very real obligation to protect the public, we must ask ourselves one of the questions that Randy asked in his letter: "Am I following through on those responsibilities I have accepted as a teacher?" Failure analysts should be the teachers for failure prevention, and our responsibilities generally include informing the public. How we follow through on this responsibility may well determine the worth of our technical contributions to society.

Journal of Failure Analysis and Prevention, 2005
Vol 5(4), p 3–4

OLD PARTS

My wife Fran and I have been "eveloping" a house for the past year. We bought the house in the 1980s, and it served wonderfully as a weekend retreat but needed substantial improvements when we decided to convert it into our retirement home. After much searching, we decided to use John Goodall of Goodall Construction: a small residential contractor who specializes in renovations and one-of-a-kind homes. As the plans evolved and the improvements developed, Fran coined the word *eveloping* to describe changes that included converting an almost flat shed roof into an *A*; adding 16 by 34 ft rooms to both stories of the house; expanding the kitchen to incorporate the front deck; installing rock veneer to the lower story, vinyl siding on the top story, new windows and doors, and a host of other things. To my surprise, the eveloping process seldom included using any of the old parts. "Perfectly good" boards, wires, sockets, piping, shelving, insulation, and other parts of the house were relegated to the garage, Habitat for Humanity, and friends' collections of things that may become useful at some future date. The contractor simply didn't want to try to incorporate old parts into an eveloping house. John convinced us that the evelopment would be cheaper, faster, and of better quality if new parts were used. He always focused on long-term rather than short-term costs: minimize maintenance, reduce upkeep, and keep future fuel costs as low as practical. Additionally, many of the old parts did not meet modern codes.

Modern codes and requirements for the small contractor are intriguing. For example, if a contractor is building a spec house, he cannot use any "reclaimed" lumber in the structure, but if he is building a contract house, the use of reclaimed lumber is acceptable. The handrails for the steps from our deck to the ground have to be gripable, while the interior handrails do not. A deck three feet off the ground

Management virtually always delights in postponing expenses, and failure analysts are frequently driven toward the short-term, low-cost solution

is required to have childproof railings, while a landscaping wall that terraces the lawn and is five feet high is not required to have any railing. The side of a house must be ten feet from a lot line, while the setback for the front is twenty feet, but decks must not count because there are no setback requirements for decking. A worker on a roof that is eight feet off the ground must be tied off, while the same worker on a deck that is ten feet off the ground has no tie-off requirements.

It is no surprise to me that the contractor did not want to use old parts. John has enough trouble keeping up with codes, standards, and requirements, and adding old parts to his work scope would only complicate his job. However, occasionally John uses an old part. I emphasize the word *occasionally.* The routine is to use new parts, but when the old part reduces short-term costs, meets codes and standards, and does not compromise long-term costs, old parts are used.

The failure analysis community frequently faces "old part" decisions. A component breaks, a system shuts down, and a failure analyst is summoned. New replacement components are difficult to find, but old parts are scattered throughout the boneyards of scavenged systems. (Our old parts include technologies that fit all too comfortably and replacement schemes that we've used before. Additionally, the "we've always done it this way" scenario applies to failure analysts as well as managers. We should never discount the benefits of experience, but too often we rely on old parts when new, improved parts are readily available.) A preliminary failure analysis of the broken component suggests that the cause of the failure was not the component but rather the service that the component received, an assembly process that misaligned the component, or some other "non-component" failure mechanism. Short-term costs are minimized by retrieving an old part from the boneyard and getting the system back in service. Management virtually always delights in postponing expenses, and failure analysts are frequently driven toward the short-term, low-cost solution: Use an old part as the replacement and get the system back in service! Quick-turnaround, low-cost solutions make the analyst look good, make the customer temporarily happy, and solve the immediate problem. After all, the part didn't cause the failure; misalignment, improper service, or some other factor was the culprit.

John's aversion to reclaimed components was a threefold concern: (a) the service that the old part might give; (b) how efficiently Joe, Kevin, Delmer, or other Goodall Construction employees could place the old part into the system; and (c) the long-term cost savings incurred by using the old part. John's experience suggests that when all three factors are considered, the use of old parts is minimized.

The lesson I learned from this old part scenario was that the long-term success of the recommendations from a failure analysis depends on numerous factors. In the hypothetical failure just outlined, the analysts must determine if the improper service was due to inherent difficulties in reaching a hidden oil fitting, if component design makes proper alignment difficult—if not impossible, and/or if the short-term savings justify the long-term costs. I'm reminded of an old story about a family that replaced their old windows with new energy-efficient windows. They were delighted with both the appearance of the windows and the reduction in their fuel bills. They were so delighted that they even wrote the company a letter expressing their appreciation. After a year, the company was surprised to find that the windows had not been paid for and called to find out why. The family replied, "We may be poor, but we aren't dumb. Your advertisement stated that the windows would pay for themselves in a year, and if this hasn't happened, don't try to blame us!"

How often do we, as failure analysts, recommend old parts, use old ideas, employ old technologies, and then try to blame others when our solutions aren't cost effective because we didn't consider the long-range implications of our recommendations?

Journal of Failure Analysis and Prevention, 2006
Vol 6(6), p 3–4

PRUNING FOR FAILURE PREVENTION

Hank Richardson is the owner and chief surgeon for The Tree Guys, a small tree service company in Pulaski, Va. He has been effective in preventing failures at 7154 Island View Way for the past several years. Hank first came to our house in response to a request for bids to remove and/or prune several trees that either blocked a view from our deck or were leaning against our boat lift. We needed Hank's expertise a second time when a tall white pine was swaying extensively in the wind and about to be blown onto our garage. Hank's third visit was to remove two trees: one that was leaning over our deck and one that was infected by pine bores. On each visit, Hank performed a visual assessment of the tree's condition, evaluated the potential damage that tree failure would produce, then estimated the remaining operating life for the tree before performing the service necessary to prevent failure. In many ways, The Tree Guys' services are similar to those performed by a failure analyst. The Tree Guys respond to three very different situations. These situations generally involve a need for either preventative maintenance, emergency services, or reclamation after a failure has occurred. As Hank discussed the causes of the potential failures in our trees, similarities between tree care and failure prevention became increasingly apparent. As failure analysts, we are most often involved with the reclamation process but should never forget the need for preventative maintenance.

The simplest services that The Tree Guys have provided are pruning, or "limbing up," healthy trees simply to improve the views of the lake, and removing a few dead or dying branches from otherwise healthy trees to improve the looks of the property and help assure against falling branches and scattered debris. These services are very similar to the maintenance activities required to assure that the workplace appearance provides a healthy image for the corporation and that the equipment works properly. Clutter, faded paint, outdated signs, and nonworking systems provide images of lack of concern, deterioration, and general difficulties within the workplace. However, the appearance of the workplace generally changes slowly, and the need for change frequently goes unnoticed until arrangements are made to host a special client or customer. Preparation for customer visits often includes the "pruning" needed to improve the looks of the workplace. As we prepared for Hank's visit, we examined our trees carefully and found pruning needs that we had not noticed prior to the examination. The pruning always improved the looks of the property. On one occasion after the trees had been pruned, a good friend asked, "What have y'all done? The place looks so much better!" The friend couldn't identify what the improvements were but did notice the change. Isn't this typical of many of the maintenance activities necessary for failure prevention? We grow accustomed to the need for paint, the increased operational noises and vibrations, and the subtle deteriorations. Preventative maintenance is easy to postpone. Unfortunately, postponement is often first recognized by our customers or, worse yet, leads to otherwise unnecessary repairs and costly services.

Hank's latest service was the removal of a locust tree that stood about three feet from our deck and towered over the house. About a year ago, we noticed that the wind swayed this tree more than either of the two locust trees nearby. The tree held our bird feeders and one end of our hammock, and we simply accepted the swaying as normal until our neighbor asked, "Have you noticed how your 'bird feeder tree' sways when the wind blows?" This question prompted us to ask Hank to examine the tree and recommend what should be done. He poked and prodded around the base of the tree, found that the tree contained a rotting section, then suggested that we remove the tree as soon as possible. Hank also checked the other locust trees for similar problems, and we were glad to hear that the other trees were healthy. I couldn't help but notice the similarity between Hank's work and the use of vibration analysis to check for impending fatigue problems.

Nondestructive evaluation (NDE) technologies have progressed to the point that most incipient failures can be detected before the component or system actually breaks. However, we must be willing to apply the technology and act on the signals received before NDE becomes helpful. The excessive swaying of the locust signaled a potential tree failure, but it took us a long time to actually act on the signal. Isn't this similar to our lack of action as we note signs of impending failures? Often we are reluctant to make, or even discuss, needed changes in our companies, organizations, and professional societies. How many failures could be avoided if we acted immediately when we received a signal that a potential problem existed? Take the time to look at your business, your community, and your societies, and prune what needs pruning, but also recognize that sometimes pruning isn't enough and there may be a "tree" that needs to be removed.

It took us about a year to be willing to accept the signals of impending failure in the locust tree. The swaying didn't appear to be dangerous, and the tree was serving several useful purposes—if bird feeders and hammocks are useful. We acted much faster when a neighbor's white pine began to lean toward our garage. Soaking rains and high winds combined to cause the pine to sway, and we could see that the roots were tearing the ground at the base of the tree. If it fell, the tree would either hit our garage or a neighbor's house. The tree needed to be removed on an emergency basis, so we called for help. Hank answered the call immediately. We moved any movable object in the tree's path—a car, a truck, and a personal watercraft—while Hank tied ropes to the tree and his truck. He took his chainsaw and climbed onto the roof of a shed that was next to the tree, then instructed the driver of The Tree Guys' truck to "tighten" as he began to notch the tree. Within ten minutes of his arrival, Hank had dropped the pine tree into our driveway without hitting anything but the driveway itself. In this case, a tree failure was barely prevented because we used a skillful tree surgeon who knew how to prune and what to do when pruning wasn't enough.

The cost of Hank's services differed for these various jobs. Pruning that could be scheduled at his convenience was much cheaper than tree removal or emergency services. Additionally, pruning, coupled with regular inspections, can prevent the need for the more expensive services. As we work with our companies, organizations, and professional societies, let's promise ourselves that we won't ignore signals of impending failures and remember that pruning and other types of preventive maintenance can be an effective and economic method for failure prevention.

Journal of Failure Analysis and Prevention, 2009
Vol 9(2), p 97–98

TUBING, BATHING SUITS, AND FAILURE PREVENTION

All my grandchildren enjoy tubing. This relatively high-speed water activity involves riding a tube that is being pulled behind a power boat at speeds approaching 20 mph. Modern tubes are not typically tubular but are semi-flat objects (ours are disc shaped) attached to the boat by a 75 ft rope. There are handles on the top surface of the tubes, and the riders, at least on our tubes, lay in a prone position while holding onto the handles. The tubes will plane on the water surface at speeds higher than about 10 mph, and if the boat is going over 15 mph, the tube will glide out of the boat's wake whenever a "sharp" curve is executed. When the water is rough and the boat's path is curvy, hanging onto the tube becomes difficult because of the bounces, and the tendency for the rider is to slide across the tube surface. The combination of centrifugal forces, slick surfaces, and bouncing makes the ride fun, but difficult. The riders can slide off the tube and into the water. The magnitude of this tendency depends on the skill and strength of the rider and the speed of the boat. The grandchildren therefore want an "awesome" ride, which requires that they be on the verge of falling off the tube during virtually every turn. As the boat driver, I have to drive so that the combinations of bumps, turns, and speed are appropriate for the rider, but, regardless of the rider's skill, awesome rides result in an occasional slip from the tube and trip into the water.

Life jackets are an essential component for any tube ride, and each of the grandchildren has their favorite jacket. On a beautiful day, the grandchildren can tube for hours without any trouble, although slips from the tubes are frequent. Visitors, however, often have problems. The problems occasionally involve a fear of sliding off the tube, or a life jacket that doesn't quite fit. However, the primary problem is the bathing suit: some bathing suits simply are not appropriate for tubing. When the rider slips into the water, the water tugs on the bathing suit and can strip the bathing suit from the falling tuber. Such stripping is primarily a problem when the bathing suit doesn't quite fit or when a female visitor wears a bikini. Over the course of this past summer, several visitors have been embarrassed because their bathing suit bottom was removed as they slipped from the tube. However, the way to avoid such embarrassment is to wear the proper equipment for tubing. This solution sounds simple, but people frequently have little desire to wear the proper bathing suit. Proper bathing suits may not be considered stylish, and the improper bathing suit may be more comfortable, and/or acquiring a new, proper bathing suit may require the investment of resources (time and money). In many ways, the bathing suit problem is similar to failure prevention in most industrial nations.

This year, easily preventable failures costing industrialized nations approximately 3% of their gross national product (GNP) will occur because people have little desire to use the proper equipment. More U.S. citizens will die in car accidents because they don't wear their seatbelts than will die as the result of enemy action in combat. Components will rust and systems will fail because no one bothers to paint the exposed part. Storage tanks and piping systems will develop leaks because failure prevention by draining the system requires the use of resources, and the addition of pitting inhibitors costs money that "can be better used to increase today's bottom line." Automobile engines will wear out and even freeze up because the oil hasn't been changed at the appropriate intervals, and radiators will become unusable because they were never flushed. Fatigue failures will occur because vibrations were ignored, and lines will become plugged because they were never cleaned. On a more personal level, breast cancers and colon cancers will continue to grow because mammograms and colonoscopies are avoided. Heart attacks and strokes will occur because the blood pressure medicine cost too much, and marriages will fail because seeking counsel is not stylish. If these personal, preventable failures are included in the cost assessment, the cost of negligence and ignorance far exceeds 3% of the GNP.

There is a tendency for us to laugh when a visitor loses a bathing suit while tubing. However, failure prevention is not a laughing matter. When a car accident results in a mangled body that requires months of rehabilitation, and the victim's injuries could have been minimal if their seat belt had been fastened, we all pay the cost of increased insurance premiums. When a pipeline breaks because preventative maintenance was not practiced, the cost of goods produced using products from that pipeline increases. And when... You get the picture! The cost of preventable failures is generally passed on to people who were completely removed from the failure process. The old adage that "I'm only hurting myself" is simply not true. We all pay for negligence, regardless of who is being negligent. In many ways, failure prevention is similar to tubing; we can act like a grandchild and wear a bathing suit that fits properly, or we can act like a visitor and everyone will have to stop and wait while we search for the lost component. Let us choose to care more about substance than style, more about performance than packaging, and more about correctness than cost. If we wisely address failure prevention, perhaps, by applying good failure prevention principles, our actions can contribute to economic recovery rather than adding to the bathing suit bottoms that are falling through the water.

Journal of Failure Analysis and Prevention, 2009
Vol 9(6), p 493–494

ALIGNING WITH THE FACTS

The United States Coast Guard Academy annually conducts an Ethics Forum. I have been privileged to speak at the Forum for the last three years. This year the privilege was enhanced by the presence of Dan Clark, the primary contributing author to the Chicken Soup for the Soul series. He was an outstanding speaker with an outstanding message—a message that pushes toward the heart of the failure analysis process. I'm not capable of reproducing Dan's message, but I am going to apply some of his ideas to the failure analysis industry.

The failure analyst is often required to provide a report, a presentation, and/or a discussion of "the evidence" and to use that evidence to sway the opinion of corporate executives, insurance agents, lawyers, judges, and jurors. Truth, or the quest for truth, should always guide the failure analysis process, but unfortunately, the failure analyst is often funded by an organization that is more interested in finding fault than finding facts. This is especially true if the fault-finding process can free the funding organization from any responsibility for the failure.

Many years ago, we were called to investigate the "quality" of a helmet that had fractured on impact with a tree. Unfortunately, a man's head was inside the helmet at the time of impact. The funding organization wanted to show that the helmet was defective, but after only minimal investigation, we concluded that there was nothing wrong with the design, material, or workmanship of the helmet. The helmet broke simply because the impact stresses exceeded the design stresses for the helmet. We phoned the funding organization, described our preliminary results, and were told to stop work. We were also told to send neither a report nor a bill and that they would bring a check to our office the next day. Additionally, someone would come by the office to collect the helmet fragments we had in our possession.

A cashier's check was delivered, but in spite of being well paid, I continue to have concerns about my participation in that failure analysis. Apparently the funding organization was not on a quest for truth, and I seriously doubt that our work ever become part of the discovery process. The facts that we learned through testing were never reported and we never heard anything further about the case. Dan Clark's ethics seminar caused me to rethink what we had done. Should we have sent a report even though we were told not to send one? Should we have asked for a check from the company rather than a cashier's check that obscured the payer? Where did our responsibility stop once we agreed to participate in the failure investigation?

Mr. Clark's message concerned beliefs, facts, and the quest for truth. We can believe something that isn't true and by selecting certain facts can distort what is true and reinforce an errant belief. Most of our beliefs are based on a combination of facts, fantasies, and illusions, and our task as failure analysts is to assure that the beliefs reflected by our reports and presentations align with the facts—all the facts—associated with the failure. Truth can best be found by aligning our beliefs with the relevant facts. The importance of the alignment can be illustrated through a fishing trip I took last fall.

My grandson Austin loves to fish. One of the first things he does when he arrives at our house is pick up a fishing pole and head for the water. He works at fishing any chance he gets, and through practice and perseverance, Austin has become a very good fisherman. He uses very light tackle and artificial lures and loves to fish the ripples in the river. On one float trip down New River, Austin hooked three very nice smallmouth bass. The first bass was lost because Austin rushed to boat the fish and broke the line in the process. The next bass was lost because he played the fish too long. He caught the third bass, which was measured and released. The bass was between 19 and 20 inches long. Those are the significant facts about the fishing trip. When we returned home, Austin was sharing his experience with his parents and said, "I should have caught three nice bass today but I only caught one. It was over 19 inches long. The other two were bigger than that but they broke my line and got away." Clearly, Austin's report was a mixture of facts and beliefs. I have no doubt that Austin believes that the two bass he lost were bigger than the one he caught. He could make a case to "prove" his position. The two bass he lost both broke his line, while the one he caught didn't break the line. Additionally, he saw all three bass in the water and his mental images are consistent with the relative sizes he presented in his report. However, his belief that the two nice bass that got away were bigger than the one he caught is primarily reinforced by his fantasies and illusions. He wants to catch nice bass. The 10- to 12-inch bass don't satisfy his fantasies, and looking at fish in the water can create an optical illusion. After all, in fishing, bigger is almost always better.

The role of the failure analyst is to develop a report that aligns the beliefs of the analyst with the facts and does not use fantasies and illusions to distort the data and analysis. In many cases, we are just like Austin: we almost have the necessary data to complete the story, but unfortunately a piece of evidence was lost. For example, in a hypothetical case, consider a situation where the fracture surface topography was almost totally destroyed by postfailure abrasion and corrosion. However, with careful examination of the fracture, limited features that could support our client's position can be seen. Unfortunately, features that oppose the client's position are also visible in a few regions. Another piece of evidence is based on a finite-element analysis that can be used to either support or refute our client's position, depending on our choice of material properties. The nominal properties for the material are tabulated in numerous handbooks and suggest that the yield strength could be just low enough to support our client's position. Unfortunately, testing demonstrated that the yield strength is actually above nominal yield, and if the actual yield is used in the analysis, our client's position is significantly weakened. A third piece of evidence fits our client's position perfectly. The question now becomes how to report the failure analysis. Do we follow Austin's example and make our report a mixture of facts, fantasies, and illusions, or do we work to assure that our report aligns with the facts and presents the truth, even if we don't have enough facts to swing the preponderance of evidence toward our client's position?

Fishermen have a reputation for stretching the truth. They let fantasies and illusions influence their beliefs, and as a result, their beliefs may not always align with the facts. Let's be certain that our clients don't think we are fishermen in the failure analysis pond, aligning our beliefs to fit the illusions and fantasies of our clients and funding agencies. As Dan Clark so clearly stated, a person's ethics can be easily seen by how well their beliefs align with all the available facts. Let's be certain that no one—not our clients, not our competitors, and certainly not our partners—believes that our reports deflect the facts so that our reported beliefs fail to reflect the total truth.

Journal of Failure Analysis and Prevention, 2007
Vol 7, p 153–154

CONSULTANTS AND CONSULTATION

Summer at the Louthans' is a time for swimming, fishing, boating, canoeing, biking, berry picking, and other outdoor activities. The grandchildren move rapidly among the various offerings and are always looking for something to add to their list of accomplishments. Without exception, they enjoy the powerboat and the thrill of being towed on skis and tubes. Most of them enjoy the challenge of a new activity and thoroughly enjoy "teaching" their friends how to fish, canoe, ride a raft, or paddle a kayak. I am amazed at how rapidly four-, five-, six-, and seven-year olds can become experts and provide free consultation to their friends. The older grandchildren are willing to advise adults, recommend techniques, and provide constructive criticism on performance, style, and attitude. They are willing to consult on virtually anything involving the waterfront: where to fish, what bait to use, how to stay on the raft during a high-speed turn, and how to hold your arms when you jump off the top of the boat dock. The grandchildren readily consult on the things they do well, and occasionally they will consult on things that they don't do at all.

This summer the grandchildren discovered an old disk in the boat shed and suddenly became experts in riding, standing, and spinning on a circular plywood platform while being towed behind the Jet Ski. Although the disk is several decades old, this summer was the first time any of my grandchildren had ever experienced a "disk ride." However, the lack of experience didn't retard the consultation. They had recommendations for the riders and for the Jet Ski drivers: "You are trying to stand too soon!" "Don't lean back when you try to spin!" "Keep your knees bent!" "Slow down!" "Speed up!" These recommendations were extrapolations from personal experiences the grandchildren had

I realized that successful consultation is generally very simple: listen to the presentations, ask questions until understanding emerges, and then share the insight that experience brings to the job.

had in related arenas. Surprisingly, those who actually listened to the advice, studied the process, and concentrated on doing what the other grandchildren recommended, learned to ride. The consultation worked, primarily because the consultants' collective voices focused the client's (rider's) attention on riding the disk.

The fundamental causes of failure of engineered systems and structures include inadequacies in design, materials selection, assembly, and service. Consultants and consultation are frequently used in efforts to minimize such failures and to maximize the potential for successful operations. Participants in the consultation process are generally highly regarded experts with either significant insight into the supporting sciences or long-term experience in operation and service. There is a belief that these outside experts can identify potential weakness in a process, program, or plan and ultimately make recommendations that lead to success. This tendency results from the in-house preparation and the consultants' determination to prove their worth. As I watched the grandchildren at the waterfront I realized that successful consultation is generally very simple: listen to the presentations, ask questions until understanding emerges, and then share the insight that experience brings to the job. The preparation required for educating the consultant—collecting and distributing the "in-house" wisdom—focuses the whole team on the project and virtually assures success.

Collectively, the grandchildren can solve almost any problem they face at the waterfront. They work as a team, and from the oldest to the youngest they take pride and joy in their accomplishments on the water. This summer, the biggest problem the grandchildren faced was learning how to use a new Super Bouncer. The instructions printed on the side of the bouncer state that the bouncer is not a trampoline and that no one should ever bounce more than two feet above the "trampoline-like" surface. After trying several locations along the water, the group attached the Super Bouncer to the front of a floating dock, and grandchildren would run across that dock, leap onto the bouncer surface, and try to use the "bounce energy" as a driver for a dive or flip into the water. On the first few tries no one was successful and the common experience was an awkward fall into the water. Some fell onto the bouncer surface and never reached the water. However, with time and continuous consultation, improvement in the bouncing process began to develop, and after a few days even the youngest could bounce into the lake with a twist, turn, or flip. The grandchildren shared observations, discussed their experiences (including failures), offered encouragement, and eventually developed a successful approach to bouncing. Statements such as "Good dive!" and "Great flip!" became common. Each child shared in the joy of the group's accomplishments.

Consultants and consultation are key resources for successful problem solving, whether the problem is the use of an old disk, a new Super Bouncer, or a complex failure analysis. The failure analyst is generally part of a team—a team that is willing to discuss, recommend, and offer advice. Hopefully, the team has the collective wisdom and includes the resources necessary to solve the problem. Consultation with this team is important and the team members should generally serve as the first consultants for most problems. Unfortunately, many of us treat our in-house failure analysis teams like my grandchildren treat their waterfront team. Consultation becomes an ad hoc process without adequate preparation and void of well-planned discussions. Certainly, success can come when we listen to the collective voices, but success is much more certain when we take the time to educate the consultants and prepare for the consultation process. I'm convinced that if we treated in-house consultants with the same formalism we generally give highly regarded outside experts, in-house consultations would become extremely successful endeavors.

The next time we are faced with the need for consultation, let's try to gather our team's wisdom through a formal, well-prepared process, rather than acting like grandchildren playing at the waterfront. Although either approach may lead to success, a well-established agenda for well-prepared team members takes advantage of the experience and wisdom available and focuses that expertise on the problem at hand. However, as we take this approach, let's not forget the simple joys of working together and sharing accomplishments.

Journal of Failure Analysis and Prevention, 2005
Vol 5(6), p 3–4

IGNORING THE INDICATORS

The average car is equipped with gages and "idiot lights" designed to warn the driver when the alternator fails to work, when the coolant temperature is too high, and/or the oil pressure is too low. Many cars have warning systems that beep to suggest that "the seat belt isn't fastened" or "the keys are in the ignition." Some cars have indicator lights to show that a door is not closed properly, and one of my cars has a light to show that the car leveling device is in operation. Lights show when the cruise control is engaged, when the emergency break is on, and when the engine needs service. There are so many indicator devices that some drivers may fail to recognize the warnings, and a few may choose to ignore the indication. However, the indicators are seldom ignored when the owner is very proud of the car or has a large personal investment (time or money) in the automobile.

All eight of my grandchildren were at our lake house for Memorial Day weekend. The eight were accompanied by six other children, thus Emily, Megan, Grant, Austin, Abby, Kate, Molly, Hunter, Jake, Ty, Ellison, Tripp, Annie Grace, and Lilly became devices that indicated the operational quality of the moment. Laughter, squeals, cries, and even silence provided reliable measures of the children's performance and of their need for attention or help. The sound indicating a fish hook in a finger differed greatly from the sound of a successful jump from the top of the boat lift into the water below. The children's mothers' ability to read the indicators amazed me. They could readily distinguish real problems from momentary concern, interpret the sounds of silence, and recognize the difference between tired, hungry, and angry cries. Whether they were sitting by the water, working in the kitchen, or talking in the living room, the mothers were tuned to the indicators from the children, especially from their own children. Not only did the mothers hear the indicators, they responded to warning signals and provided the service required. The weekend passed without a single child being lost, abandoned, or seriously hurt. The fish hook in a finger was the most serious incident. I think that everyone, including the mothers, had a good time, partially because someone paid attention to the performance indicators.

Many industrial systems have performance indicators such as vibration sensors, temperature indicating devices, corrosion monitors, and/or other on-line process sensors. Unfortunately, the signals for such monitors are occasionally ignored. I was recently involved with a failure where the signal from an "operating" thermocouple was assumed to be in error and was therefore ignored because a "back-up" thermocouple indicated the anticipated operating temperature. However, the back-up thermocouple was improperly positioned and continued operation caused system meltdown. Many of us have investigated failures that occurred because someone ignored an indication. The ignored indication may have come from an on-line monitor, been an unusual sound from the machinery, or been provided by a change in system performance. Regardless of the source of the indication, someone chose to ignore its signal and failure occurred.

My wife Fran is tuned to the indicators in our cars. She notices squeaks that I'd never hear if she didn't call them to my attention. She notices subtle changes in the ride and/or the acceleration. In spite of the fact that she can't see the gas gage, she notices when the "change oil" or "service engine soon" light comes on. She not only notices these indicators, she responds to them and has the car properly serviced. In fact, since our children are grown and, along with the grandchildren, live hundreds of miles from our home, Fran mothers our cars. Consequently, most of our automobiles go several hundred thousand miles before they wear out.

Mothers don't ignore the indicators from their children. Furthermore, they realize that children regularly require service, that proper service improves performance, and a lack of service may cause failure. I recognize that sometimes mothers are overly attentive to service and will do silly things such as homework and other duties that should be required of their children, but the point is that mothers respond to the indicator signals from their children. Unfortunately, engineers, scientists, foremen, operators, technicians, and other responsible parties (male and female alike) may ignore the indicators from their industrial operating systems, especially if they have little personal investment (ideas, time, career, or money) in the system.

The differences between mothering and operating an industrial system are significant, but we could learn something about successful operations by examining mothering. Mothers take pride in performance, have concern about health, and drive toward success. Mothers care about the future—tomorrow is as important as today. Mothers will take the time required to respond to the indicators and will hassle others to be sure that their children are serviced properly. Seldom does a mother say, "My child is crying. Will you take care of her?" Most mothers will offer comfort themselves before they involve someone else in the service process. We all have learned a great deal from our mothers and most of us know that we could improve our own performance and the performance of the components and systems under our control if we, like our mothers, listened to the indicators, learned to interpret the signals, and hastened to respond when response was needed. When we ignore a signal we are actually saying, "I don't care about that system, and, since I don't care, I'll ignore the indicator because I have something more important to do such as drink a cup of coffee or eat a donut."

Let's all make the effort required to provide the proper response to the indicators we monitor. If we do, our future may hold fewer failures to investigate and more successes to praise.

Journal of Failure Analysis and Prevention, 2004
Vol 4(4), p 3–4

LISTENING

My brother Barnes and I were in a canoe, drifting down scenic New River in Virginia, fishing for smallmouth bass and catching very few. The lack of activity caused our conversation to move from fishing to a variety of other topics. Barnes, who is nine years younger than I and retired five years before I did, was a very successful biologist/salesman before his retirement, with customers throughout the world. As we rounded a bend in the river, a small For Sale sign prompted me to ask, "What is the secret to success as a salesman?" Barnes' immediate reply surprised me. Without any hesitation he said, "Listening," and then explained to me why true listening is very hard to accomplish.

People, especially educated people, don't listen very well because they don't listen very long. In any conversation among people from Western nations, the person "listening" only listens until something that is said creates an opportunity for a response. Once that opportunity is created, the "listener" is now thinking about what they are going to say, rather than focusing on what the other person is saying. To illustrate his point, Barnes related this personal experience.

Sales were declining, so Barnes called a meeting of his international sales group to obtain the sales prospective on how to reverse the trend. Lots of ideas were presented, tabulated, and discussed. Charts appeared on the wall and I assume that the discussions conformed to today's understanding that, for fear of hurting someone's feelings, nothing bad is said about any of the ideas. The modern approach to creativity is to conclude that no idea can be considered bad, regardless of how bad the idea is in reality. However, a very lengthy discussion concluded that competition is tough and market shares are hard to hold. Barnes stopped the discussion and said, "I think the problem is that we don't listen to the customer. We need to practice listening."

The indignant and defensive reaction of the salesmen was that they did listen to the customer. With this reaction, Barnes suggested that the meeting move to other considerations, and said, "Let's consider our presentations, especially our spontaneous presentations. To practice this, I want each of you to consider the best vacation that you have ever had, and in a few minutes I'll select several of you to make a presentation on the vacation. But before we do that, I want to read you a poem:

> Two small fish, standing in a line,
> Three big bears, feeling just fine,
> Four fat turkeys, wallowing in mud,
> Five small seals, covered with blood.
> Six sneaky rats, climbing the wall,
> Seven alligators, starting to fall,
> Eight lazy llamas, lying on dirt,
> Nine small hands, wiping on shirt.
> Ten stunned salesmen, trying to recall
> Words of a poem, heard by all."

Before any discussion could develop, Barnes gave a test to each of the salesmen. The questions on the test were similar to:

- What were the alligators doing?
- How many seals were present?
- What animals were feeling fine?

The test had ten questions. Not one salesman got all ten correct, and several missed at least half the questions. I actually don't remember many of the questions because my attention was diverted from Barnes' discussion to a fish that struck the lure I was casting into the river. After all, we were fishing and it is easy not to listen when there is a bass on your line.

What if you had attended Barnes' meeting? Would you have been listening to the poem or would you have been thinking about the response you would make about your vacation? How would you have done on the test? How well do you listen to your customers, and what distractions keep you from hearing what your customer is saying? Distractions that keep us from listening come in many forms.

I've been distracted while hearing many presentations because I had a concern about something the author said during the introduction and spent the next five minutes formulating the question I was going to ask. I've been distracted because I spoke or responded to someone sitting next to me. I've been distracted because I was the next speaker. I've listened to presentations looking for the mistakes rather than looking for the pearls of wisdom. But most of all, I've been distracted when I should have been listening because something was said that opened the door for me to speak. Once the door was opened, I thought about what I was going to say rather than what was being said.

As failure analysts we need to listen, and listen carefully. All too often something important is overlooked when we fail to listen. As you collect background information concerning a failure, do you listen, or do you ask questions to demonstrate your knowledge of the subject? When a customer is presenting a multitude of concerns, do you listen to all the concerns, or do you begin to generate a response to the first or second concern presented?

Listening! I think my little brother is correct. He attributes his success to his ability to listen, and perhaps each of us could be professionally more successful if we would learn to listen better. However, even if listening doesn't lead us to greater professional successes, it can improve our lives. Our spouses, children, parents, friends, and fellow workers would appreciate our willingness to listen. It doesn't sound hard, but listening is perhaps the most difficult thing we do, and an increased ability to listen will make each of us better associates, and better people.

Journal of Failure Analysis and Prevention, 2009
Vol 9, p 183–184

OBJECTIVITY: IS IT NECESSARY?

Several decades ago, my wife Fran was discussing child rearing with a seasoned lady who was experiencing difficulty with one of her adult children. At that time we had only two children, whose ages were five and one. As with most preschoolers, the children were relatively easy to control; Fran said "do" and the children "did." At least they gave the impression that they were trying to accomplish the tasks she had assigned, and, if they got too far out of hand, the children's attitudes could readily be adjusted. Fran felt that the seasoned lady was spending entirely too much time trying to get her son out of jail and suggested that a week or two behind bars would do the boy some good. Perhaps the lady simply needed to leave the boy alone. "Let him reap his just rewards." Fran has never forgotten the lady's reply: "Once a mother, always a mother, and the bigger the child, the harder he pulls on a mother's heartstrings." Since that time, Fran has learned the magnificent and sometimes frightening truth behind those words. Being a mother is a continuous, never-ending job that must be approached with love, wisdom, knowledge, resourcefulness, and long-term commitment. In many respects, a successful failure analysis must be approached in the same fashion.

Several years ago I had the opportunity to participate in a workshop that examined the potential for microbial influences on a group of corrosion-induced failures. There were two distinct camps at the workshop. One camp could find evidence for microbial influences in virtually all the failures, while the other camp could find little evidence for any microbial effects on any of the failures. Clearly, the camps saw the same things in a totally different fashion. The microbial-influenced corrosion (MIC) camp consisted primarily of engineers and scientists who had published extensively in the MIC arena, while the conventional corrosion camp had seldom, if ever, mentioned MIC in any publication. Apparently, the ability to "see" the influences of microbes on corrosion processes increased as the expertise in the field increased. It was almost as if the MIC camp were populated with "mothers" and MIC technology was their favorite child. Although we never settled on the exact role of microbes in the various failures, the discussions did suggest that the non-MIC camp was overlooking microbial effects that were obvious to the MIC camp. It is hard to overlook contributions from your own children, thus it was easy for the MIC camp to see the potential for microbial effects.

The "children" from the MIC camp are not unique. Generally, if a technology begins to pull on a technologist's heartstrings, that technology becomes the technologist's child. I once heard a graduate student who was examining a fracture surface state, "I could prove that this failure occurred by virtually any process; there is evidence for overload, stress-corrosion cracking, hydrogen embrittlement, and fatigue. Which one do you like best?" Unfortunately, his comments were very appropriate. There was evidence for lots of failure processes and the cause of failure was difficult to define. This is not unusual, and the cause(s) of many failures cannot be established with total certainty. We must deal with the preponderance of evidence.

My heartstrings resonate to discussions of the effects of the environment on the behavior of engineering materials. Hydrogen embrittlement, stress corrosion, corrosion fatigue, irradiation damage, and liquid metal embrittlement somehow capture my imagination. Whether browsing in the library, attending a conference, or simply discussing metallurgy/materials engineering, I'll read about, listen to, and/or discuss environmentally induced failures in preference to almost any other metallurgical subject. When I see a failure, I feel a strong desire to understand the role that environmentally induced degradation may have played in the failure process. But, even though I'm not necessarily objective, I do recognize that environmental effects are not the sole cause of failures.

I have a friend whose heartstrings are pulled by design-induced problems and another friend who believes that mechanics considerations are the primary keys to any failure analysis. The three of us come from different technical backgrounds, have different expertise, and approach failure analysis in far different fashions. Over the years we've learned to listen to each other and not place a huge priority on our personal preferences. However, although none of us is really objective and each has a favorite "child," I believe that all three of us are good failure analysts.

Any good analyst recognizes the importance of looking at a failure from several sides, the necessity of determining the most probable cause by collecting data, evaluating hypotheses, and testing theories. The good analyst is also cognizant of the requirement to distinguish findings based on demonstrated fact from findings based on conjecture and/or hope. Initially, I felt that the processes of collecting, evaluating, testing, and distinguishing should establish objectivity and minimize the tendency for an analyst to be led by a heartstring pull. However, as the editorial evolved it became apparent to me that following the pull of your heartstring is important. As a technology and/or failure process becomes a favorite "child", following that "child" will lead to understanding, competency, and expertise. When expertise is developed, the failure analyst, like any other parent, may tend to focus on the "child's" contribution. However, the good failure analyst will also recognize that other "children" are important and will be continually aware of personal bias.

Is objectivity really necessary to a successful failure analysis? A dictionary definition of being objective is to be "without bias or prejudice; detached; impersonal." Perhaps holding an objective position is not only unnecessary; objectivity may not even be suitable for an experienced failure analyst. Experiences produce knowledge and knowledge produces biases. Furthermore, can one actually conduct a successful failure analysis if he or she remains detached and not personally involved? I don't think so! The failure analyst must be somewhat like that seasoned mother—personally involved and willing to commit the time and energy necessary to ensure success, even when the failure process seems jailed by the data and information available.

Journal of Failure Analysis and Prevention, 2004
Vol 4(1), p 3–4

REVISION TOWARD EXCELLENCE

McGuire's Family Campground is less than a mile from Exit 101 on I-81 near Dublin, Va. Two of my friends, Mike and Sherry McGuire, are the owner/operators of the campground, its restaurant, and a Saturday evening dance known as the "Hillbilly Opry." The campground is about a mile from Claytor Lake State Park and is typically filled with RVs, campers, and regular guests. Sherry serves a selection of wonderful breakfasts, a hamburger that will tickle your taste buds and cause you to gain weight, and home-cooked dinners with some of the best desserts in Southwest Virginia. My wife Fran and I try to eat at McGuire's at least once a week when we are in Virginia. Although many of the customers are regulars, there are no strangers, and table-to-table conversation is promoted by the friendly bantering of Mike and Sherry and their adult children, Jason and Jennifer, who take orders, pour coffee, cook, and bring the food. If you are stopping by for the first time, try to get there at about 6:30 on a Saturday evening. This will allow time for a home-cooked meal before the Hillbilly Opry starts at 7:30. With Mike on the banjo, Jason on the bass, Wade Petty on the fiddle, and Gordon Cox on the guitar, Southern Heart will play some of the best bluegrass and country music you'll ever hear. A hundred or so guests will listen, flatfoot, line dance, and/or slow dance, depending on the music, displaying their abilities and a willingness to learn. The Opry is a family affair with children, parents, and grandparents frequently dancing to the same song but not necessarily to the same beat. When the music stops at about 10:30 p.m., the dinner, dance, and fellowship will have cost a couple no more than $30 if they had dessert, extra cups of coffee at the dance, and invested a dollar in the half-and-half drawing. If the dollar investment pays off, the winner's share will cover the cost of the evening.

I fancy myself a poet and frequently use poetry to express feelings to my family. When my mother-in-law died this past January, I wrote a poem for Fran. The poem was a series of verses that covered our lives from dating through our 42 years of marriage. Each set of verse ended with the phrase "she's the best that's ever been." I asked Mike to put the words to music and have Southern Heart perform the song at the Hillbilly Opry on the Saturday before Memorial Day. We were having our annual fishing and barbecue gathering and were planning to top off the day by attending the dance. I thought that a surprise performance of her poem would be special for Fran. A week before Memorial Day, Mike said, "I want you to hear the song before next Saturday. I've changed the words around and need to be sure that you like it." I was shocked. How could Mike rearrange the words? The poem was mine and I had asked for music, not for advice on the poetry. However, when Mike and Jason played and sang the revised edition, I realized that Mike had transformed my poetry into a wonderful love story. Perhaps the transformation is illustrated best by the change of one word in the ending phrase. He replaced the word *she's* with the word *you're*. He made the song personal: "You're the best that's ever been." When Jason sang the song at our private presentation, the tears in Fran's eyes revealed the meaning it held for her. That wasn't the only change Mike made. He even wrote a new verse that focused on the future. All of his changes were improvements. Mike reviewed my poem, read my feelings, and expressed them perfectly. Mike and Jason publicly debuted the song on May 24, and the prolonged applause and continuous compliments attested to Mike's musical ability, the quality of his review and advice, and the excellence that resulted. Was I ever pleased and proud!

One function of the editorial review board for *Practical Failure Analysis* is to review manuscripts and offer advice to authors. Accepting advice is frequently difficult, especially when the advice is unsolicited, unexpected, and, often, unwanted. A recommendation for change often shocks authors, but when they realize that by accepting the advice, they can transform the manuscript, improve the technical position, and/or correct mistakes, many authors use the review as an occasion to revise toward excellence. After receipt of the reviewer's comments, one author wrote, "I must say that, at first, I was a bit put out by the introduction of the concept of hydrogen-assisted cracking as the underlying crack mechanism, as opposed to SCC as set forth in my analysis of the fracture. My research into the matter since I received the reviewer's comments and edits did, however, prove them to be correct. In this regard, I am grateful for their care and concern." This comment demonstrates three stages of feelings that are common whenever unanticipated advice is offered and shows how the review process can help create excellence.

We are initially "put out" when the advice of a reviewer is to change the manuscript. How could the reviewer question our approach or our conclusions? After all, aren't we the experts? Didn't we do the work? The "put out" may even include anger. On rare occasions, authors have withdrawn manuscripts rather than accept or even challenge a reviewer's comment. Humility and discipline are required to move from the "put out" stage to the "acceptance stage." Humility is required to accept that there *might* be a better pathway, interpretation, or approach, and discipline is required to place emphasis on the word *might*. Free advice is often only worth its cost; however, reviewer's comments are seldom free. Consider the time required for reading, understanding, and commenting on a manuscript. The best news reviewers can receive from an author is "My research into the matter ... did, however, prove them to be correct." The author transitioned from the "put out" stage into the "acceptance" stage because he investigated the reviewer's comments. After finding that the comments were correct, the author accepted the advice, altered the manuscript, and improved the article. After the improvements were made, the author entered the "grateful" stage of feelings. He recognized that acceptance of the reviewer's advice avoided an error and ultimately increased his satisfaction with, and the excellence of, the publication. However, effort and a willingness to listen to advice are required for an author to move into the "grateful" stage.

Many times during the course of an analysis, study, or investigation, colleagues will offer "free" advice. Well-planned programs will be in progress when the advice suggests revisions and alterations. It is generally easy to discount or ignore this advice, especially if it comes from friends who were asked to confirm that the investigative pathways were appropriate. Confirmation is so much easier to accept than program-altering advice, but, when heeded, the program-altering advice can transform the average into the excellent. Revision toward excellence generally requires cooperation, effort, and, above all, a careful review by a highly qualified reviewer.

As I reflect on Mike and Jason's performance of "The Best That's Ever Been," I'm beginning to understand fully the "grateful" stage of accepting advice. Mike took my poem and built it into an excellent song. He added the music, and the song became something that Fran and I will treasure always. I'm grateful for that, truly grateful. The fourth time I heard the song sung, someone in the audience shouted, "Take it to Nashville!" just after the music stopped. We may do that, but regardless of where the song goes, it was transformed because of changes recommended by a friend, reviewer, and expert in his field. Mike built on what I gave him and the product was far more than I could have achieved on my own. How often do we dance to the wrong tune, or even sit out the dance entirely, simply because we refuse to accept advice or, worse yet, take advice without investigating what the new tune may bring to the occasion?

Practical Failure Analysis, 2003, Vol 3(5), p 3–4

THE LIST

My wife Fran occasionally allows me to go to the grocery store by myself. This is an infrequent occurrence because experience has demonstrated that I often purchase things that we do not need, such as premium ice cream, freshly baked pies, and aged angus beef steaks. Additionally, the very thing that I went to the store to purchase is often forgotten. Because of my malpractices, Fran now gives me a list to carry to the store and encourages me to use the list while I am in the store. There are many kinds of lists: shopping lists, Christmas card lists, lists of telephone numbers, e-mail addresses, and birthdays; however, the list I most often use is a shopping list for the grocery store. Because the grocery list has proven useful, I have evaluated some of the subtleness that Fran includes.

The list may say lettuce, cucumbers, and carrots. It may say frozen baby limas, frozen mixed vegetables, and cherry pie filling. Yellow cake mix, boneless chicken breasts, and canned green chilies may be included, but the list never says soap. We use soap! Both of us bathe regularly with soap and water, but the list never says soap. The list always says Dial Soap. Fran does not want just any soap—she wants a specific brand of soap: Dial. The list will also never say black-eyed peas; it will say Bush's Black-Eyed Peas. There are several other items that are always listed under a brand name. Generally, a list of 15 to 20 items will only contain two or three brand names. The brand names provide an assurance of taste, composition, or quality that may not be associated with another brand. Fran's list is based on personal taste, experience, and years of success with a specific brand and an occasional culinary disaster that resulted from using a different brand. Infrequently, she will write, "Do not get Brand X," next to an item on the list.

The failure analysis community also has a list—actually, several lists. The lists are not generally available in written form, but virtually all of us know about them. There are lists of companies that provide needed services in a timely fashion and at a reasonable price. There are lists of companies that seem to always have an excuse for not meeting expectations. Potential bidders that will typically complete a job in time and under budget are on one list, while another list includes companies that tend to provide a low bid and then come back several times asking for more money to complete the task. Although these lists are seldom published, they are available through phone calls, discussions with friends, and informal e-mails. The names on the list change from time to time, and after only one misstep, a name from the good performance list may move down. Two or more poor performances will virtually assure removal from the good list. Movement from the poor list to the good list is much harder. Rich McNitt, retired head of the Engineering Science and Mechanics Department at Penn State, once said that in the eyes of most faculty members, the good department head was always only one mistake away from needing to be replaced, while the poor department head needed seven significant accomplishments, similar to walking on water, before becoming marginally adequate. Although the list is fluid, upward movement is much more difficult than falling, because most of us are reluctant to change an experienced-based, bad opinion.

Realization that "the list" exists and is informally shared among professional friends and associates should invoke the questions "What list am I on?" and "How do my actions impact my company's position on the list?" It is extremely difficult to become the brand name of choice, especially because everyone does not have the same taste, needs, and experience. However, as we make our list of professional accomplishments we should achieve today, do we place a priority on the difficult, or do we fill the list with the easy and leave the difficult to someone else in the company? "Brand name" people tend to attempt the difficult. There is even an old adage that "the difficult we do immediately; the impossible takes a little longer." Ultimately, our actions demonstrate where we want our company to be listed.

Fran and I recently saw the motion picture "The Bucket List," which described two dying men and their actions to accomplish a list of things that they wanted to do before they kicked the bucket (died). The bucket list becomes more personal when you are retired and over sixty-five, because the time for accomplishment is somewhat limited. Most of us have a mental bucket list and our personal priorities reveal that list. However, as working failure analysts, the real question does not concern the items we have on our own list but concerns how we are listed by our customers and clients. How does our performance measure up to the quality our customers deserve? Are we a brand name, a good generic company, or are we down there with the poor performers?

The list that I carry to the grocery store is very similar to the list that many corporations have concerning failure analysis and metallurgical services. There are lots of generic offerings, but there are places on the list reserved for the Dial Soaps and Bush's Black-Eyed Peas. Most of us desire to have our reputation and performance put us in those special places. After all, none of us wants to be Brand X. So as the next job arrives, remember Rich McNitt's observation: It takes seven outstanding performances to overcome one lackluster job.

Journal of Failure Analysis and Prevention, 2008
Vol 8, p 213–214